Table of (

Chromatin Immunoprecipitation Essentials

By

Lenin Dev

A Practical Guide to DNA-Protein Interaction Analysis

Published by: Phoenix Wings Press LLC
Contact: phoenixwingspress@gmail.com

About the Author

Lenin Dev is a molecular biologist with a strong focus on gene regulation and DNA-protein interactions. With years of practical lab experience, he specializes in making complex techniques like Chromatin Immunoprecipitation (ChIP) easy to understand and apply.

Passionate about science education, Lenin is committed to helping students and researchers gain clear, hands-on knowledge of essential molecular biology tools. This book reflects his goal to simplify ChIP and support the next generation of scientists.

Acknowledgments

To the mentors who taught me, the students who inspired me, and the colleagues who shared their wisdom—thank you. I am deeply grateful to every hand that contributed to a ChIP experiment and every mind that challenged the boundaries of understanding.

To my family and friends, your patience, encouragement, and unwavering belief made this possible.

And to the global scientific community—this book is my humble tribute to your curiosity, dedication, and courage in exploring the unknown.

1. INTRODUCTION TO CHROMATIN AND GENE REGULATION

In the heart of every human cell lies a silent symphony of life—an intricately wound string of genetic code that holds the instructions for all biological functions. This code, the DNA, does not float freely within the nucleus. Instead, it exists in a complex, dynamic partnership with specialized proteins known as histones. Together, they form chromatin—a beautifully engineered structure that is at once protective, compact, and astonishingly responsive. Understanding chromatin is akin to decoding the language of cellular potential. Within this scaffold, the secrets of gene expression, regulation, identity, and adaptability reside. And within the folds of chromatin, we begin our journey.

At first glance, the human genome appears to be a long, uninterrupted string of nucleotides—over three billion base pairs in length—capable of encoding around 20,000 genes. But DNA alone is not enough to dictate the full complexity of cellular behavior. It must be interpreted, accessed, and controlled. This is where chromatin plays its pivotal role. Acting as both guardian and gatekeeper, chromatin determines when, where, and to what extent genetic instructions are read.

It is no passive container; it is an active participant in the governance of gene expression.

The structure of chromatin is hierarchical and elegant. The fundamental unit is the nucleosome: a spool-like core of eight histone proteins—two each of H2A, H2B, H3, and H4—around which approximately 147 base pairs of DNA wrap nearly two turns. Linker DNA connects one nucleosome to the next, forming what resembles beads on a string. This "beads-on-a-string" structure is the least compacted state of chromatin and is often associated with regions of active transcription. It is also the most accessible conformation for DNA-binding proteins. However, the story does not end here.

To fit into the tiny nuclear space, this string must undergo further compaction. Nucleosomes coil into higher-order structures, which can fold and loop to achieve a densely packed state. Depending on the degree of this compaction, chromatin is broadly categorized into two forms: euchromatin and heterochromatin. Euchromatin is loosely packed, transcriptionally active, and rich in genes. It is the site where the majority of gene expression occurs, and it remains relatively accessible to transcription factors and RNA polymerase. In contrast, heterochromatin is tightly compacted, gene-poor, and generally transcriptionally silent. Its primary function is structural, helping maintain chromosome integrity, regulate gene silencing, and protect repetitive DNA regions.

This dynamic between euchromatin and heterochromatin is not fixed; rather, it is fluid and subject to constant regulation. Cells adjust chromatin architecture in response to developmental cues, environmental stimuli, and internal signaling pathways. But how does the cell orchestrate such a finely tuned performance? The answer lies in epigenetic modifications—chemical tags that decorate histone proteins and the DNA itself, influencing chromatin behavior without altering the underlying genetic sequence.

Among the most well-studied epigenetic marks are histone acetylation and methylation. Acetylation typically correlates with transcriptional activation. When acetyl groups are added to the tails of histones, the positive charge of these proteins is neutralized. This weakens the interaction between histones and the negatively charged DNA, leading to a more relaxed chromatin structure. In this open state, transcriptional machinery can more easily access DNA. Histone acetyltransferases (HATs) are the enzymes responsible for this modification, and their counterparts, histone deacetylases (HDACs), remove these acetyl groups to repress gene activity.

Methylation, on the other hand, is more nuanced. Depending on the specific amino acid residue modified and the number of methyl groups added, histone methylation can either activate or repress transcription. For instance, trimethylation of histone H3 at lysine 4 (H3K4me3) is a hallmark of active promoters, whereas H3K27me3 is a classic marker of repressed genes. These marks are not random; they serve as docking sites for reader proteins that recruit additional regulators, chromatin remodelers, or even RNA polymerases themselves.

But histones are not the only molecules subject to epigenetic modification. DNA itself can be methylated, particularly at cytosine bases in CpG dinucleotides. DNA methylation is strongly associated with long-term gene silencing and plays an essential role in processes like

X-chromosome inactivation, genomic imprinting, and suppression of transposable elements. In cancer biology, aberrant DNA methylation patterns can lead to inappropriate activation or silencing of critical genes, contributing to tumorigenesis.

With this complex code of epigenetic regulation in place, the cell has developed a robust system to fine-tune gene expression. Still, how do scientists decode this layered information? How can we visualize, map, and quantify the interactions between DNA and the proteins that regulate its expression? This is where chromatin immunoprecipitation—or ChIP—emerges as a transformative tool in molecular biology.

ChIP allows researchers to capture a snapshot of DNA-protein interactions in vivo. By using antibodies that target specific proteins or modifications, scientists can pull down chromatin fragments that are bound to these factors. Once isolated, the DNA can be analyzed to determine which genomic regions are involved. This technique bridges structure and function, showing where transcription factors bind, where histone modifications are enriched, and how these patterns change over time or in response to treatments.

But before one can fully grasp the power of ChIP, it's crucial to understand the vocabulary of chromatin biology. Words like "enhancer," "promoter," "transcription factor," "insulator," and "nucleosome positioning" are not mere jargon—they are coordinates in the genomic landscape. Enhancers are distal elements that boost transcription from a distance, often looping through three-dimensional space to contact promoters. Promoters are regions immediately upstream of genes where transcription begins. Transcription factors are the interpreters—proteins that read the language of the genome and transmit instructions to the transcriptional machinery. Insulators block the spread of chromatin states, acting as boundaries between active and repressive domains. And nucleosome positioning determines whether a gene is buried or exposed, ready to be transcribed or sentenced to silence.

Visualizing these concepts transforms abstract mechanisms into concrete understanding. Imagine a diagram of chromatin remodeling—enzymes sliding nucleosomes along DNA or evicting them entirely to expose regulatory sequences. Picture the interplay of histone marks, like a molecular barcode read by effector proteins. Or consider how mutations in chromatin regulators, such as those seen in developmental disorders or cancers, alter the epigenetic code and derail normal gene expression programs.

Chromatin is not a static shell; it is responsive, malleable, and deeply contextual. The same stretch of DNA can be silent in one cell and active in another, depending on the chromatin environment. This flexibility underlies cell differentiation, tissue identity, and organismal complexity. It is also the source of phenotypic diversity, even among genetically identical individuals.

In exploring the chromatin landscape, we begin to see the genome not as a rigid script, but as a dynamic score—one that can be interpreted in countless ways depending on the players involved. The same sequence can generate vastly different biological outcomes. This epigenetic layer of control enables organisms to adapt, specialize, and evolve. It is here, at the intersection of structure and function, that molecular biology reveals its most poetic truths.

As we close this introductory chapter, one thing becomes clear: the regulation of gene expression is not governed by DNA sequence alone. It is shaped by the packaging, accessibility, and decoration of chromatin. These layers of regulation are not peripheral—they are essential. They define how genes are turned on or off, when they are expressed, and to what extent. Without chromatin, there would be no orchestration, no temporal control, no context-specific expression. Life would be static. Instead, thanks to chromatin, life is beautifully dynamic.

In the chapters ahead, we will explore how chromatin structure is interrogated experimentally. We will dissect the methodology of ChIP, trace its evolution, and uncover its vast applications—from developmental biology to disease research. But none of that is possible without a firm understanding of the foundation we have just laid.

The nucleus is more than a vault of DNA. It is a living archive, constantly rewritten, remodeled, and reinterpreted by the language of chromatin.

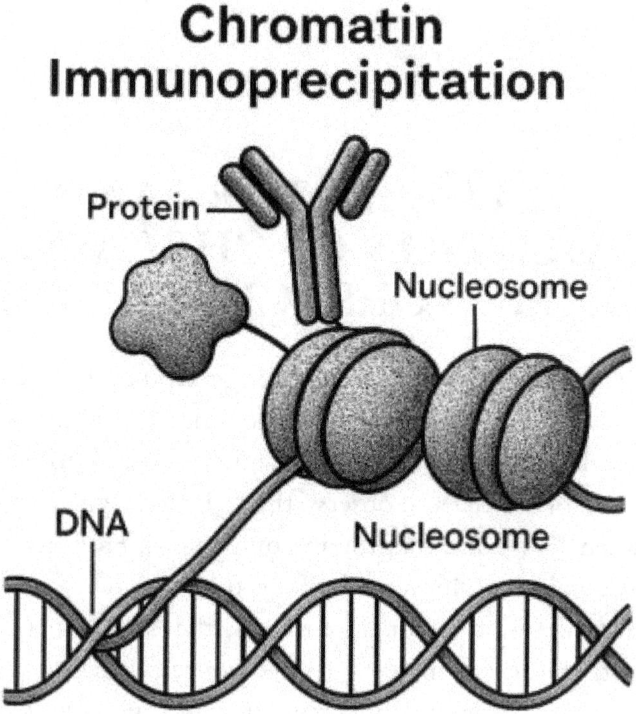
Chromatin
Immunoprecipitation
Protein
Nucleosome
DNA
Nucleosome

2. THE HISTORY AND EVOLUTION OF CHROMATIN IMMUNOPRECIPITATION

The story of chromatin immunoprecipitation, or ChIP, is not merely the tale of a laboratory technique—it is a chronicle of humanity's evolving ability to peer into the inner sanctum of the cell and observe the intimate, regulatory dance between DNA and its protein companions. From its earliest conceptual roots in yeast biology to its current status as a cornerstone of functional genomics, the evolution of ChIP mirrors the rise of molecular biology itself. It began humbly, born out of a desire to understand how histones—long known as structural components of chromatin—could influence gene expression, and it matured into one of the most transformative methodologies in biological research.

In the earliest days of chromatin research, DNA was understood largely through the lens of its sequence, the infamous four-letter alphabet that spelled out the code of life. But as the 20th century wore on, scientists began to suspect that more was at play—specifically, that gene expression could be governed not only by the sequence itself but by the accessibility and regulation of that sequence. This brought attention to chromatin: a tightly packed, protein-coated form of DNA whose structural complexity hinted at layers of control far beyond what the naked sequence could reveal. The histone proteins, around which DNA coiled to form nucleosomes, were once thought to be passive packaging materials. However, early observations suggested they might in fact play active roles in gene regulation. Scientists began searching for methods that could physically capture interactions between specific proteins and the DNA sequences they bound.

It was in the 1980s that the foundational concept of immunoprecipitating protein-DNA complexes began to crystallize. The principle was straightforward in theory but ambitious in execution: could one capture, isolate, and identify the precise DNA regions bound by specific proteins in living cells? Could this be done without destroying the native interactions within chromatin? Initial efforts involved the use of formaldehyde to "freeze" protein-DNA interactions in place—a process known as crosslinking—so that even after cells were broken open, these interactions would remain stable. This was a leap forward. Crosslinking preserved the in vivo landscape of chromatin, enabling the recovery of protein-bound DNA fragments with greater confidence.

What followed was a technical progression that would shape the fate of ChIP as a mainstay of molecular biology. After crosslinking, the next hurdle was to shear the DNA into manageable fragments while preserving the crosslinked complexes. Sonication—using high-frequency sound waves to physically shear DNA—emerged as the most consistent method. It enabled researchers to reduce chromatin into small, reproducible fragments without compromising the protein-DNA bonds preserved by crosslinking. From here, antibodies came into play. By using antibodies specific to a protein of interest, scientists could selectively pull down the associated DNA regions. The method had all the makings of a revolution: it could map where transcription factors, histone modifications, or other chromatin-associated proteins were acting within the genome.

Yet early ChIP experiments were painstaking and low-throughput. Researchers would perform ChIP, then identify the pulled-down DNA using PCR or Southern blotting—methods that could only examine a few loci at a time. Nevertheless, even these initial applications yielded profound insights. In yeast, scientists used ChIP to demonstrate that certain histone modifications were enriched at actively transcribed genes, directly tying chromatin structure to gene activity. These findings would reverberate through the field and set the stage for broader, genome-wide approaches.

The next leap came with the integration of ChIP with microarray technology, giving rise to ChIP-chip. In this method, DNA fragments isolated via ChIP were fluorescently labeled and hybridized to genomic microarrays—slides printed with probes corresponding to known DNA sequences across the genome. This allowed researchers to map protein-DNA interactions across large genomic regions in a single experiment. Suddenly, ChIP was no longer limited to individual genes; it could probe regulatory landscapes on a much broader scale. The technique was powerful, but not without limitations. Microarrays required prior knowledge of genome sequences, suffered from hybridization biases, and offered limited resolution. But the concept of genome-wide mapping had taken root.

Then, in the mid-2000s, came the true renaissance. With the rise of next-generation sequencing technologies, ChIP found a new partner: deep sequencing. This marriage birthed ChIP-seq, an approach that abandoned microarrays in favor of sequencing every DNA fragment pulled down by immunoprecipitation. The advantages were immediate and profound. ChIP-seq offered unparalleled resolution, dynamic range, and genome coverage. It eliminated the constraints of pre-designed arrays, allowing researchers to explore entire genomes—including non-coding regions and previously unannotated elements—with unprecedented clarity.

ChIP-seq did more than improve data quality; it reshaped how scientists asked questions. For the first time, they could build comprehensive maps of transcription factor binding sites, chromatin modifications, and regulatory elements across entire genomes in different cell types, developmental stages, and disease states. Landmark papers soon followed. In 2007, the ENCODE project began using ChIP-seq to map histone modifications, RNA polymerase occupancy, and transcription factor binding across the human genome, revealing a level of regulatory complexity that far exceeded prior expectations. These studies unveiled enhancer landscapes, chromatin states, and long-range chromatin interactions, laying the groundwork for modern epigenomics.

The success of ChIP-seq, however, depended on continued innovation in methodology. Improvements in antibody specificity became a major focus. Early ChIP experiments often suffered from off-target binding and low signal-to-noise ratios.

The development of rigorously validated antibodies for specific histone modifications and transcription factors enhanced the reproducibility and reliability of results. In parallel, refinements in crosslinking protocols, chromatin shearing techniques, and library preparation methods contributed to the robustness of ChIP-seq workflows.

Yet, as the popularity of ChIP-seq grew, so did its challenges. It required large numbers of cells, often in the millions, making it unsuitable for rare populations or clinical samples. The signal could be noisy, and background artifacts introduced by non-specific antibody binding or over-crosslinking complicated data interpretation. Moreover, the method's reliance on crosslinking and fragmentation raised concerns about artificial enrichment or loss of transient interactions.

These limitations fueled the next wave of technological evolution. In recent years, several innovations have emerged to address the inherent constraints of traditional ChIP. One of the most transformative has been the development of single-cell ChIP technologies. These methods aim to profile protein-DNA interactions in individual cells, capturing heterogeneity that is otherwise masked in bulk analyses. Though technically demanding and still under active refinement, single-cell ChIP represents a bold step toward understanding chromatin regulation at the level of individual cellular identities.

Another breakthrough came in the form of CUT&Tag (Cleavage Under Targets and Tagmentation), a method that eliminates the need for crosslinking, sonication, and complex immunoprecipitation. Instead, it uses a fusion of protein A and Tn5 transposase, directed to specific chromatin proteins by antibodies. Upon binding, the enzyme cuts and inserts sequencing adapters directly at the site of interaction. This streamlined workflow dramatically reduces input requirements and background noise while enhancing resolution. CUT&Tag has quickly gained traction as a next-generation alternative to ChIP, particularly in low-input or high-throughput applications.

Throughout this journey, the field has been shaped by visionary scientists and seminal publications. Researchers such as Michael Grunstein, David Allis, and Brad Bernstein made pivotal contributions to our understanding of chromatin biology and the development of ChIP-based tools. Their work illuminated the functional consequences of histone modifications, demonstrated the importance of chromatin remodeling in development and disease, and set methodological standards that continue to guide the field today.

To understand the trajectory of ChIP is to witness a microcosm of scientific progress—curiosity turned into methodology, methodology into insight, and insight into transformation. From its humble beginnings in yeast chromatin studies to its modern reincarnations in single-cell and enzymatic profiling, ChIP has remained at the forefront of molecular exploration. It has not only opened windows into the hidden architecture of the genome but has provided a template for how technology and biology can evolve in tandem.

As we continue to innovate, the spirit of ChIP endures. It is a method, yes, but also a mindset: a belief that the secrets of gene regulation can be brought to light with the right tools, the right questions, and the right persistence. The chapter of ChIP is still being written, now in finer resolution, deeper layers, and broader applications than ever before.

And so, as we close this chapter on the history of ChIP, we do not mark an ending but a transition. The technique has matured, diversified, and inspired an entire generation of molecular biologists. What began as a fragile, hopeful protocol has become an engine of discovery—one that continues to reveal the subtle orchestration of the genome, note by note, interaction by interaction.

3. UNDERSTANDING CHROMATIN STRUCTURE AND NUCLEOSOME DYNAMICS

Within the microscopic confines of the cell nucleus, an extraordinary structure unfolds—one that balances the need for both order and flexibility. Chromatin, the physical form in which our DNA exists in vivo, is more than a packaging solution. It is a highly organized, dynamic framework that dictates the functional potential of the genome. At the heart of this architecture lies the nucleosome, the fundamental repeating unit of chromatin. Understanding how nucleosomes are positioned, modified, and mobilized reveals not just the mechanics of DNA compaction but the very basis of gene regulation itself.

Each nucleosome consists of approximately 147 base pairs of DNA wrapped nearly two turns around a histone octamer. This octamer is composed of two copies each of the core histones: H2A, H2B, H3, and H4. These histones are not simply inert spools; they are biochemical regulators, equipped with flexible amino-terminal tails that extend outward and serve as sites for a variety of post-translational modifications. Through this structure, DNA is compacted roughly sixfold, enabling the enormous human genome to fit within the tiny confines of the nucleus. But compaction is only one part of the story. The position and density of these nucleosomes along the genome play a critical role in determining which regions of DNA are accessible to transcription factors, RNA polymerase, and other regulatory machinery.

Nucleosome positioning is not random. In fact, cells deploy complex regulatory systems to ensure that nucleosomes are placed in ways that support, rather than hinder, transcriptional programs. At the promoters of actively transcribed genes, for instance, there often exists a nucleosome-depleted region—a stretch of DNA that remains accessible to transcription initiation complexes. Flanking this open region are well-positioned nucleosomes, arranged in a phased manner, providing both structural support and regulatory control. Conversely, genes that are silenced tend to be densely packed with nucleosomes, creating a physical barrier to the transcriptional machinery.

The dynamics of nucleosome positioning are profoundly influenced by the incorporation of histone variants—non-canonical histone proteins that replace standard histones within the nucleosome. These variants alter the biochemical and structural properties of the nucleosome, influencing its stability, DNA binding affinity, and interactions with chromatin-associated factors. For example, the histone variant H2A.Z is frequently found near gene promoters and enhancers, where it facilitates chromatin opening and transcriptional activation. Another variant, macroH2A, is associated with gene repression and the establishment of silent chromatin domains. The selective deposition of these variants is a tightly regulated process, often orchestrated by specialized histone chaperones that recognize specific genomic regions and cellular cues.

Yet the mere presence of nucleosomes and histone variants is not enough to account for the remarkable plasticity of chromatin. Cells require mechanisms to move, evict, or restructure nucleosomes in real time in response to environmental or developmental signals. This is where ATP-dependent chromatin remodeling complexes come into play. These multi-subunit machines harness the energy of ATP hydrolysis to disrupt histone-DNA contacts, enabling nucleosomes to slide along DNA, be ejected entirely, or be replaced with variants. Remodeling complexes are diverse in composition and function, falling into families such as SWI/SNF, ISWI, CHD, and INO80. Each plays unique roles in transcription, DNA repair, replication, and recombination.

Consider the SWI/SNF complex—among the best characterized of these remodelers. It functions by repositioning nucleosomes to expose gene regulatory elements. In doing so, it facilitates transcription factor binding and gene activation. In contrast, the ISWI family is known for maintaining nucleosome spacing and repression. The dynamic interplay between these remodelers ensures that chromatin remains both stable and responsive, capable of swift transformation in the face of changing cellular conditions.

The density of nucleosomes across a gene's regulatory region is another key determinant of transcriptional accessibility. A densely packed nucleosome array can occlude transcription factor binding sites, thereby repressing gene expression. Conversely, reduced nucleosome density in regulatory regions increases the likelihood that transcription factors will locate and bind their target motifs. This principle is particularly important in understanding how transcriptional regulation operates in a context-dependent manner. The same DNA sequence may be completely silent in one cell type but highly active in another, simply due to differences in nucleosome architecture and chromatin remodeling activity.

To understand these relationships empirically, scientists have developed a range of techniques to map nucleosome occupancy and positioning across the genome. These include micrococcal nuclease digestion followed by sequencing (MNase-seq), which identifies regions of protected DNA corresponding to nucleosome-bound fragments, and more recent techniques like ATAC-seq, which uses a hyperactive transposase to probe regions of open chromatin. These methods have revealed that nucleosomes are not evenly spaced throughout the genome. Instead, they are strategically positioned to reflect the functional needs of the cell, creating a chromatin landscape that guides and modulates gene expression programs.

Even more fascinating is how nucleosomes can be repositioned or disrupted during active gene transcription. During the initiation phase, RNA polymerase and associated factors require access to promoter regions, necessitating nucleosome eviction or repositioning. Once transcription is underway, elongating polymerases must navigate through the chromatin fiber, temporarily displacing nucleosomes and recruiting histone chaperones to reassemble them behind the transcriptional machinery. This cyclical process of disruption and reassembly is vital for maintaining chromatin integrity while enabling gene expression.

The movement of nucleosomes during gene activation is not random; it follows a tightly choreographed sequence of events. Diagrams illustrating this process reveal a dance of displacement and return—one nucleosome shifting aside as RNA polymerase approaches, another settling back into place once transcription has passed. This dynamic remodeling ensures that the genome remains both accessible and protected, balancing the need for responsiveness with the necessity of genomic stability.

The functional implications of chromatin dynamics are vast. Gene silencing, for instance, can result from the establishment of nucleosome arrays that obscure promoter elements and inhibit the binding of activators. Conversely, gene activation often requires nucleosome remodeling to expose enhancer and promoter regions, allowing transcription factors to bind. These opposing chromatin states are not fixed; they are continually remodeled in response to signals ranging from hormonal cues to stress responses. The plasticity of chromatin structure thus allows cells to reprogram gene expression profiles in real time, supporting rapid adaptation and phenotypic diversity.

This dynamism has particular relevance in the context of chromatin immunoprecipitation (ChIP). The technique depends on capturing protein-DNA interactions as they exist in vivo. Regions of open chromatin, where nucleosomes are sparse or highly mobile, are more likely to allow transcription factors and regulatory proteins to bind. As a result, these regions often yield strong signals in ChIP experiments. In contrast, tightly packed heterochromatic regions may be less accessible to proteins and, therefore, less likely to show enrichment in ChIP assays. This means that chromatin state and nucleosome dynamics directly influence the landscape of protein-DNA interactions that ChIP is designed to uncover.

Understanding the nuances of chromatin organization and nucleosome behavior allows researchers to better interpret ChIP data and appreciate the broader regulatory architecture of the genome. The epigenetic language written in nucleosome positions, histone modifications, and remodeling patterns is a crucial determinant of when and where genes are expressed. It reveals how seemingly identical sequences can yield vastly different functional outcomes depending on the chromatin environment in which they reside.

As we uncover more about the factors that shape chromatin structure—from ATP-dependent remodelers to histone chaperones, from nucleosome spacing to variant incorporation—we deepen our understanding of cellular identity, gene regulation, and developmental potential. Chromatin is not merely a scaffold; it is a living, breathing interface between the genome and the world around it. Its responsiveness to internal and external stimuli makes it one of the most fascinating and consequential components of molecular biology.

The study of chromatin dynamics is, in many ways, the study of cellular decision-making. Every repositioned nucleosome, every histone mark, every remodeled segment of DNA tells a story about what the cell is doing, what it remembers, and what it is preparing for. In this ever-shifting landscape, nucleosomes are both sentinels and storytellers—guarding the genome while narrating its unfolding chapters.

4. DNA-PROTEIN INTERACTIONS

The genome, while encoded in a four-letter alphabet, does not speak for itself. Its instructions, though fundamental, are inert without the molecular interpreters that bind, read, and regulate its content. Among these interpreters are a remarkable class of molecules known as DNA-binding proteins. These include transcription factors, repressors, architectural proteins, and chromatin modifiers—each playing a nuanced role in the grand orchestration of gene expression. They are the conductors of the genome's symphony, dictating when a gene must be turned on, when it must be silenced, and under what conditions it can be allowed to whisper or shout.

At the heart of this regulation lies the specific recognition of DNA motifs. Despite the seemingly uniform structure of the DNA double helix, the sequence imparts subtle variations in shape, flexibility, and electrostatic potential that make specific regions attractive docking sites for particular proteins. Transcription factors are equipped with specialized structural motifs that allow them to latch onto these sequences with both strength and specificity. These interactions are not random; they are guided by evolutionary constraints that have honed both the DNA sequence motifs and the protein domains that recognize them.

Zinc fingers represent one of the most versatile DNA-binding motifs in eukaryotes. These structures use one or more zinc ions coordinated by conserved cysteine and histidine residues to stabilize their fold. The classic C2H2 zinc finger, for instance, consists of a short β-sheet and an α-helix, with the helix inserting into the major groove of the DNA. By combining multiple zinc fingers in tandem, a single protein can recognize extended DNA sequences, greatly enhancing specificity. This modular nature also makes zinc finger proteins attractive tools for genome editing and synthetic biology.

Leucine zippers, another widespread motif, operate through a different mechanism. These proteins form dimers—either homo- or heterodimers—through hydrophobic interactions between leucine residues spaced every seventh position along an α-helix. Once dimerized, their basic regions flank the zipper and bind to DNA, often at palindromic sequences. The bZIP family of transcription factors, including AP-1 and CREB, relies on this motif to regulate genes involved in growth, differentiation, and stress responses.

Then there are helix-turn-helix motifs, first discovered in bacterial repressors and later found in eukaryotic homeodomain proteins. This motif features two α-helices joined by a short turn, with the recognition helix fitting snugly into the major groove of DNA. The precision with which these motifs recognize DNA is extraordinary; a single amino acid substitution in the recognition helix can drastically alter binding specificity, underscoring the finely tuned nature of DNA-protein interactions.

But proteins rarely act in isolation. In the complex milieu of the nucleus, DNA-binding proteins often function as part of larger assemblies—multi-protein complexes that integrate signals, amplify effects, and exert sophisticated regulatory control. One such example is the Mediator complex, a colossal molecular machine that bridges sequence-specific transcription factors at enhancers and promoters with the basal transcription machinery at the core promoter. By doing so, Mediator acts as an information hub, integrating diverse signals from multiple transcription factors to fine-tune the initiation of transcription by RNA polymerase II.

Equally important, but functionally opposed, is the Polycomb repressive complex (PRC). This multi-subunit group of proteins is central to the maintenance of gene silencing through chromatin-based mechanisms. PRC2 catalyzes the trimethylation of histone H3 at lysine 27 (H3K27me3), a mark associated with repressed chromatin. This modification then recruits PRC1, which compacts chromatin and prevents transcriptional activation. Polycomb complexes are crucial in development, where they maintain the silent state of genes that should only be expressed in specific tissues or at particular times.

The idea that a single transcription factor binds a single gene and turns it on or off is a simplification that underrepresents the complexity of gene regulation. In reality, most genes are regulated by multiple transcription factors acting in concert—a phenomenon known as combinatorial control. This allows for an enormous diversity of gene expression profiles using a limited set of regulatory proteins. The same transcription factor may activate one gene while repressing another, depending on the presence of co-factors, chromatin context, and post-translational modifications. This combinatorial nature permits subtle gradations of control, akin to dimming a light rather than simply flipping a switch.

To understand this at a functional level, consider the regulation of the *IFNG* gene in immune cells. Its expression is governed by the combined actions of T-bet, STAT4, and NFAT transcription factors, each binding to distinct elements within the promoter and enhancer regions. These interactions do not occur simultaneously or in a vacuum; they are coordinated in response to cellular signaling pathways. The integration of these signals ultimately determines whether the *IFNG* gene is transcribed, to what extent, and for how long. This dynamic control ensures a precise immune response, preventing both overactivation and inadequate defense.

Another illustrative example lies in the regulation of developmental genes like *HOX* clusters. These genes, which dictate body plan and segment identity, are tightly regulated through enhancers bound by specific transcription factors and silencers recruited by Polycomb proteins. Their expression patterns are spatially and temporally defined—activated in one region of the embryo and repressed in another. This spatial control is vital for proper tissue patterning, and disruptions can lead to congenital malformations or developmental disorders.

These examples underscore a fundamental truth: gene regulation is not merely a product of static DNA elements but of dynamic, context-dependent DNA-protein interactions that respond to a constellation of biological inputs. Timing, location, signal strength, and chromatin context all converge to shape the transcriptional output. A gene may be "poised" for activation, bearing the molecular hallmarks of readiness, yet remain silent until the appropriate combination of factors is present. This poised state is often observed in stem cells, where developmental genes are held in check by bivalent chromatin marks—H3K4me3 (active) and H3K27me3 (repressive)—until the cell commits to a lineage.

Unraveling these complex interactions is a formidable challenge, one that requires tools capable of capturing protein-DNA contacts in their native cellular environment. This is precisely where chromatin immunoprecipitation (ChIP) proves indispensable. By using antibodies directed against specific transcription factors or chromatin-associated proteins, ChIP allows scientists to "freeze" and extract the genomic regions occupied by these factors. When combined with sequencing (ChIP-seq), the technique provides a genome-wide map of binding sites, offering critical insights into how regulatory proteins shape gene expression patterns.

The ability to visualize these binding events in living cells has revolutionized our understanding of transcriptional control. We now know, for instance, that enhancers can be located tens or even hundreds of kilobases away from the genes they regulate, looping through three-dimensional nuclear space to make contact with target promoters. These interactions are mediated by architectural proteins such as CTCF and cohesin, which help establish topologically associating domains (TADs) that constrain enhancer-promoter communication within defined boundaries. Such insights would have remained elusive without the spatial resolution offered by ChIP and related technologies.

Even more compelling is the discovery that many disease-associated genetic variants identified by genome-wide association studies (GWAS) do not lie within protein-coding regions but within non-coding regulatory elements—enhancers, silencers, and insulators. Understanding which transcription factors bind to these regions, and how that binding changes in different cellular states or disease contexts, is crucial for decoding the functional consequences of these variants. ChIP-seq has played a central role in mapping these interactions and linking genetic variation to altered gene regulation.

Temporal dynamics add yet another layer of complexity. Transcription factor binding is not static; it fluctuates in response to signaling cascades, circadian rhythms, and developmental transitions. A factor that binds strongly during cell division may vanish entirely during quiescence, only to return under stress or hormonal stimulation. These fluctuations are tightly regulated, often involving post-translational modifications such as phosphorylation, acetylation, or ubiquitination that alter the protein's affinity for DNA or its ability to recruit co-regulators.

This temporal modulation is particularly evident in the context of inducible genes—those activated rapidly in response to external stimuli. Consider the *FOS* gene, an immediate early gene activated by MAPK signaling. Within minutes of stimulation, transcription factors such as Elk-1 and SRF bind the *FOS* promoter, recruiting co-activators and RNA polymerase II to initiate transcription. ChIP experiments reveal this sequence of events in real time, capturing the waves of protein recruitment that precede and accompany gene activation.

Such detailed regulatory choreography reinforces the concept that DNA-protein interactions are at the core of cellular identity and function. They define the boundaries between silence and speech in the genome, between potential and action. To read these interactions is to read the regulatory code—a language written in motifs, translated by domains, interpreted by complexes, and modified by time and space.

For researchers seeking to decode this language, mastery of ChIP is not optional—it is essential. The ability to track where, when, and how transcription factors engage the genome transforms abstract models of gene regulation into empirical maps. These maps are not merely academic; they guide the development of therapeutics, inform the diagnosis of diseases, and illuminate the evolutionary pressures that have shaped our genome.

As we deepen our understanding of DNA-protein interactions, we begin to appreciate the exquisite precision of cellular control systems. What may seem chaotic from afar is, in fact, an ordered network of signals, motifs, and machines working in harmony. Each protein, each motif, each complex contributes a note to the symphony, conducted by the rules of biochemistry and refined by the needs of the organism.

And in this symphony, the regulatory code is not static—it is alive. It adapts, evolves, and remembers. It is written not just in DNA, but in the interactions that sculpt its expression. It is a code that cannot be cracked by sequence alone, but one that opens itself through the study of structure, function, and interaction. In decoding it, we move closer to understanding not just how life is built, but how it learns, adapts, and thrives.

5. OVERVIEW OF CHIP AND ITS SCIENTIFIC IMPACT

In the intricate choreography of cellular function, nothing is left to chance. Every gene, every regulatory element, every fleeting molecular interaction unfolds with surgical precision to maintain the vitality of life. At the core of this symphony lies the ability of proteins to engage with DNA at specific loci, guiding the fate of transcriptional activity and, ultimately, cellular behavior. Yet, until a few decades ago, these interactions remained largely invisible—hidden within the dense nucleus, unreachable by conventional molecular tools. Then came chromatin immunoprecipitation, or ChIP, a breakthrough that would forever alter the landscape of molecular biology.

Chromatin immunoprecipitation offered more than a method—it offered a window. A window into how transcription factors found their targets, how histone marks organized epigenetic memory, and how enhancers whispered across vast genomic distances to activate genes in precisely timed sequences. It allowed scientists not merely to infer but to observe. Suddenly, protein-DNA interactions that had once existed only as abstract ideas could be captured, sequenced, and mapped with astonishing resolution.

At its core, ChIP is an elegant biochemical strategy. The process begins with a living cell, frozen in time through crosslinking agents that preserve protein-DNA interactions as they naturally exist within the nucleus. The chromatin is then fragmented—either by enzymatic digestion or through mechanical sonication—into manageable pieces. An antibody, chosen for its specificity to the protein of interest, is introduced into this complex mixture. Through immunoprecipitation, this antibody isolates the fragments of DNA bound by the targeted protein. These fragments, once purified and de-crosslinked, reveal a molecular footprint: a tangible record of the genome segments engaged by regulatory machinery at a given moment in time.

What began as a technique for histone modification mapping in yeast has since evolved into one of the most versatile and powerful tools in modern molecular biology. The reagents that make this method work—crosslinkers like formaldehyde, lysis buffers optimized to preserve chromatin integrity, magnetic or agarose beads to anchor the antibody complexes—each contribute to the precision and reliability of the assay. Yet, it is the antibody that remains the linchpin of success. A well-characterized, highly specific antibody determines whether ChIP will yield a clean, interpretable profile or be marred by background noise and ambiguity.

As ChIP matured, so too did its applications. What began in small-scale experiments quickly expanded to genome-wide investigations with the rise of high-throughput sequencing. ChIP-seq emerged as a revolutionary fusion of chromatin biochemistry and next-generation sequencing, enabling researchers to map the binding sites of transcription factors, histone modifications, and chromatin-associated proteins across entire genomes with remarkable sensitivity. This capability turned ChIP into more than a tool—it became a lens through which the regulatory architecture of the cell could be studied in both healthy and diseased states.

Perhaps one of the most transformative impacts of ChIP has been in oncology. Cancer, at its core, is a disease of deregulated gene expression, often driven by aberrant transcription factor activity and epigenetic disruption. ChIP-seq has enabled the discovery of oncogenic enhancers, transcriptional addiction networks, and reprogrammed epigenomes across tumor types. In stem cell biology, ChIP has illuminated how pluripotency factors like Oct4, Sox2, and Nanog coordinate to maintain self-renewal while priming cells for lineage-specific differentiation. In immunology, ChIP has shed light on the regulation of cytokine genes, enhancer dynamics during immune activation, and the epigenetic underpinnings of memory T-cell formation.

Despite its power, ChIP is not without limitations. The assay is technically demanding, often requiring careful optimization of fixation times, chromatin shearing conditions, and antibody validation. Low-abundance proteins may yield weak signals. Some genomic regions are inherently resistant to fragmentation, leading to coverage gaps. Moreover, antibody cross-reactivity or poor affinity can mislead interpretation. These challenges underscore the importance of rigorous controls and validation strategies. Nonetheless, when executed with precision, ChIP reveals patterns of regulation that would otherwise remain concealed within the chromatin landscape.

One of ChIP's greatest strengths lies in its ability to operate in the context of the native nucleus. Unlike in vitro assays, which may not accurately reflect chromatin state or nuclear architecture, ChIP captures protein-DNA interactions as they exist within living cells. This in vivo relevance makes ChIP indispensable for understanding gene regulation in real biological systems. It is particularly useful for identifying transcription factor binding sites within enhancers, promoters, and other regulatory elements. In addition, ChIP enables the mapping of histone modifications—chemical marks that define chromatin states such as activation, repression, and bivalency.

As researchers began to apply ChIP across more cell types, conditions, and developmental stages, the technique expanded beyond individual proteins to encompass networks of regulatory factors. It became clear that the genome is not regulated in isolation but through the combinatorial actions of multiple proteins acting in concert. ChIP helped define super-enhancers, regions of unusually high transcription factor occupancy and chromatin activity that drive expression of key identity genes. These insights gave rise to new models of transcriptional regulation, where chromatin landscapes are dynamically remodeled in response to developmental cues and environmental stimuli.

The scientific impact of ChIP also extends into the realm of systems biology. By integrating ChIP data with transcriptomics, epigenomics, and three-dimensional chromatin conformation maps, researchers can now construct multilayered regulatory networks. These models help decode how cells make fate decisions, respond to signaling pathways, or transition between states such as quiescence, proliferation, and senescence. ChIP has become a foundational technique for consortiums such as ENCODE and the Roadmap Epigenomics Project, where its data has helped annotate the noncoding genome and refine functional maps of human regulatory elements.

The classical ChIP workflow, although subject to continual refinement, remains conceptually straightforward: crosslink, lyse, shear, immunoprecipitate, purify, and detect. The addition of high-throughput sequencing or microarrays transforms these immunoprecipitated fragments into global binding maps. Bioinformatics pipelines then align reads to reference genomes, call peaks of enrichment, and assign putative target genes. This experimental and analytical journey—from chromatin to sequencing to interpretation—has revolutionized how we understand transcriptional control and epigenetic regulation.

As with all scientific techniques, evolution is inevitable. The field has now begun embracing newer iterations of ChIP designed to overcome traditional limitations. Variants such as ChIP-exo, ChIPmentation, and CUT&Tag offer increased resolution, reduced input requirements, or streamlined protocols. Single-cell ChIP technologies are emerging to explore regulatory heterogeneity across individual cells, particularly within complex tissues or developing embryos. These innovations continue to build on the foundational principles of ChIP while adapting to the demands of modern genomics.

As readers will discover in the chapters ahead, mastering the theory and practice of ChIP is not merely a technical accomplishment—it is a gateway to understanding the regulatory logic of the genome. It is through this lens that gene regulation ceases to be an abstraction and becomes a decipherable, mappable code. ChIP has not only enabled this transformation; it has driven it.

6. CHOOSING THE RIGHT MODEL SYSTEM

In the intricate world of chromatin biology, the power of chromatin immunoprecipitation—commonly known as ChIP—relies not just on technical precision but on an often-overlooked element: the choice of biological model. As with any powerful tool, its true potential is only unlocked when wielded with context. That context begins with the living system in which chromatin dynamics are studied. Whether exploring the regulation of pluripotency in embryonic stem cells or tracking oncogene activation in aggressive tumor tissues, the biological canvas on which ChIP is applied shapes every outcome, every signal, and every insight that emerges from the assay.

The decision of whether to use immortalized cell lines, freshly isolated primary cells, or intact tissue samples is never trivial. It dictates not only the feasibility of the experiment but the biological relevance of the results. Each model system carries a unique fingerprint of gene expression, chromatin accessibility, and nuclear architecture. Immortalized cell lines such as HeLa, K562, or HEK293 have long been the workhorses of molecular biology due to their ease of culture and scalability. These lines offer predictable growth kinetics and abundant material for repeated experimentation. Yet they carry the burden of genetic drift and chromatin landscapes that may be far removed from their tissue of origin. As a result, transcription factor binding or histone modification patterns captured in these cells may not reflect physiological conditions.

Primary cells, on the other hand, provide a more faithful snapshot of in vivo biology. Isolated directly from tissues or blood, they retain much of their native epigenetic landscape. For ChIP applications focused on immune cell activation, tissue-specific gene regulation, or patient-derived pathologies, primary cells offer unmatched relevance. However, their scarcity, variability between donors, and limited proliferative capacity introduce challenges in achieving sufficient chromatin input and experimental replicates. Crosslinking and lysis conditions must be meticulously optimized to avoid loss of signal or excessive chromatin shearing.

Then there is the world of tissue samples—frozen biopsies, formalin-fixed paraffin-embedded sections, and even laser-microdissected regions—each providing a glimpse into gene regulation in its architectural and microenvironmental context. Tissues preserve cell-cell interactions, extracellular matrix influence, and spatial heterogeneity that may drive differential chromatin states. Performing ChIP on such complex matrices, however, requires not only technical finesse but also consideration of accessibility and signal-to-noise ratios. Tissue fixation protocols, enzymatic digestion methods, and nuclei isolation steps must be calibrated with care, often uniquely tailored for each tissue type.

The question of model system expands beyond the human context. In model organisms like *Saccharomyces cerevisiae*, the simplicity of the genome and the power of genetic tools have made yeast a foundational system for studying histone modifications and transcriptional regulation. In *Drosophila melanogaster*, researchers have mapped Polycomb response elements and developmental enhancers using ChIP. The mouse remains a gold standard for in vivo studies, offering genetically engineered lines and disease models that closely recapitulate human conditions. Human cell lines, though inherently artificial, still dominate due to their accessibility and the wealth of genomic data available. Yet each organism brings not only biological insights but technical parameters: genome size, repeat content, nuclear structure, and chromatin compaction—all of which influence ChIP performance.

Developmental stage is another pivotal variable. The same transcription factor may bind distinct genomic loci in embryonic versus adult cells, reflecting changes in chromatin accessibility and co-factor availability. Epigenetic states are fluid during differentiation, and ChIP performed at different developmental windows can reveal how gene regulatory networks are sequentially activated or silenced. For instance, pluripotency factors like Oct4 and Sox2 occupy a different chromatin landscape in stem cells than in differentiated neurons. Thus, ChIP not only captures static snapshots but can also chronicle the epigenomic choreography across time.

In disease models, especially in oncology and neurodegeneration, the chromatin landscape often bears the signature of transformation or degeneration. Tumor samples may harbor mutations in chromatin modifiers, altered enhancer usage, or aberrant silencing of tumor suppressor genes. ChIP in these contexts unveils how miswired epigenetic programs drive pathology. Patient-derived xenografts, organoids, and CRISPR-modified cell lines represent emerging models that blend relevance with manipulability. Yet ethical considerations—particularly in human studies—must be addressed. Informed consent, de-identification, and institutional oversight govern the use of patient samples for ChIP experiments, especially when sequencing data could reveal incidental genetic information.

In recent years, the emergence of single-cell technologies has shattered the illusion of cellular homogeneity. Bulk ChIP, by design, averages signals across millions of cells, masking rare regulatory events or subpopulation-specific binding. The advent of single-cell ChIP and derivative methods like CUT&RUN and CUT&Tag has allowed researchers to explore chromatin landscapes in rare populations such as cancer stem cells, circulating tumor cells, or early embryonic blastomeres. These methods, while still maturing, demand new model systems and new ways of thinking. When every cell tells a different story, the challenge shifts from signal amplification to data interpretation.

Purifying or enriching specific cell types becomes critical when working within heterogeneous samples. Fluorescence-activated cell sorting (FACS) allows precise selection of cells based on surface markers, fluorescent reporters, or even cell cycle phase. Magnetic bead separation, though less precise, offers high-throughput enrichment for well-characterized populations like CD4+ T cells or CD34+ stem cells. These strategies not only enhance signal fidelity in ChIP but also reduce confounding background from irrelevant cell types. In studies involving the brain, for instance, isolating neurons from glial cells prior to ChIP can unmask regulatory elements specific to cognition or memory formation.

Beyond biology and technique, the design of proper controls and replicates lies at the heart of reliable ChIP data. Input DNA serves as a baseline, representing the genome-wide distribution of sonicated chromatin. Mock IPs with isotype control antibodies help detect non-specific enrichment. Biological replicates ensure that observed patterns are not artifacts of stochastic variability or sample processing. And technical replicates, while not mandatory in every case, help calibrate assay reproducibility and sequencing depth requirements.

Each model system, from yeast to human organoids, from bulk tissues to single cells, offers a unique vantage point on the complex interplay between DNA and the proteins that interpret it. The choice of model is therefore not a footnote in the experimental design but a foundational decision—one that determines the scope, sensitivity, and significance of ChIP-derived insights. Just as a microscope's resolution is governed by its optics, the clarity of ChIP results is shaped by the biology from which they are drawn. Choosing wisely allows the technology to illuminate gene regulation with precision; choosing poorly renders even the best protocol impotent.

And as we march toward a future of integrated omics, personalized medicine, and cellular atlases, the responsibility of selecting the right model system becomes even greater. It is no longer enough to ask what the genome contains—we must ask where, when, and in whom these features are active. ChIP, empowered by thoughtfully chosen models, continues to be our flashlight in this darkened maze of regulatory complexity, helping us uncover how life's blueprint is read, interpreted, and ultimately, rewritten in health and disease.

7. SAMPLE PREPARATION AND CROSSLINKING STRATEGIES

Behind every successful chromatin immunoprecipitation experiment lies an invisible architecture of precision—one that starts long before antibodies are added or DNA fragments are pulled down. That structure is built during sample preparation, the quiet yet critical stage that determines whether your ChIP assay will yield meaningful insights or disappear into background noise. In the choreography of chromatin and protein capture, the way you treat your biological material before immunoprecipitation dictates everything that follows. It is here, in the earliest steps, where the epigenome can be either faithfully preserved or irreversibly lost.

Crosslinking, though deceptively simple, is the pivot point of this preparation. It preserves protein-DNA interactions at the moment they exist in the cell, freezing a dynamic world into place long enough for molecular interrogation. The most widely used crosslinker in ChIP is formaldehyde—a small, permeable molecule that forms reversible covalent bonds between amino and imino groups. It acts fast, usually within 10 minutes, creating methylene bridges that tether transcription factors, histones, and chromatin-associated proteins to DNA. Formaldehyde's appeal lies in this balance: strong enough to capture interactions, but gentle enough to allow reversal for DNA recovery.

Yet formaldehyde is not the only player in the crosslinking game. In many systems, especially when targeting proteins that do not directly bind DNA but are part of larger protein complexes, a single crosslinker may not be sufficient. Enter DSG (disuccinimidyl glutarate) and EGS (ethylene glycol bis[succinimidyl succinate])—longer, amine-reactive, homobifunctional crosslinkers that can capture protein-protein interactions with greater spatial flexibility. When used in combination with formaldehyde, they extend the capture radius, locking in indirect chromatin associations that formaldehyde alone might miss.

However, with added power comes added risk. Over-crosslinking is a common pitfall that can severely reduce antibody accessibility to epitopes and hinder chromatin fragmentation. The more extensively a chromatin fiber is crosslinked, the more resistant it becomes to sonication or enzymatic digestion. Worse still, excessive crosslinking can trap cellular debris and cause irreversible DNA damage, which compromises the integrity of downstream sequencing. Conversely, under-crosslinking risks the loss of transient or weak interactions and may result in chromatin that shears too easily or dissociates prematurely, yielding false negatives.

Optimization, therefore, becomes a dance—a fine-tuned balancing act between fixation time, temperature, and the nature of your sample. For cultured cell lines, a 1% formaldehyde solution applied for 8–10 minutes at room temperature is typically sufficient. Primary cells may require slightly shorter exposure, especially if they are small or delicate.

Tissue samples, being denser and less permeable, often require perfusion or agitation to allow even penetration of the crosslinker. In such cases, protocols may extend fixation to 15 or 20 minutes, but only under careful monitoring. Post-fixation, quenching with glycine halts the crosslinking process, preventing further bond formation and preserving cellular morphology.

Speed is everything. From the moment cells are harvested to the final quenching step, time becomes a fragile resource. Cold temperatures and rapid processing reduce proteolytic degradation, preserve chromatin architecture, and maintain protein epitopes in a state recognizable to antibodies. This is why most ChIP protocols are performed on ice or at 4°C, with protease and phosphatase inhibitors included in every buffer. These inhibitors protect the integrity of transcription factors, chromatin remodelers, and histone variants, many of which are rapidly cleaved or modified post-lysis if unprotected.

Nuclease inhibitors such as sodium butyrate or EDTA are equally important, especially when preparing samples for histone modification ChIP. Endogenous nucleases, released during cell lysis, can fragment DNA prematurely and bias the resulting chromatin pool. Inhibitors serve as sentinels—holding back the tide of degradation long enough for ChIP to lock in the data. A well-stocked lysis buffer becomes a molecular fortress: defending the chromatin from internal sabotage.

Once crosslinked and quenched, the next challenge lies in breaking the cellular barrier and accessing the chromatin without shattering it. The process of cell lysis and chromatin extraction is deceptively complex, particularly in samples with tough cell walls or dense extracellular matrices. For most mammalian cells, a two-step lysis protocol suffices. The first buffer disrupts the plasma membrane, releasing the nuclei. The second buffer, often supplemented with detergents like SDS, NP-40, or Triton X-100, solubilizes the nuclear envelope, exposing the chromatin to sonication or enzymatic digestion.

Sonication remains the gold standard for fragmenting chromatin in ChIP. High-frequency ultrasound waves shear DNA to an average size between 200 and 500 base pairs, ideal for immunoprecipitation and sequencing. Yet sonication is both art and science. Over-sonication can fragment epitopes or denature proteins, while under-sonication yields high-molecular-weight DNA that hampers resolution. Each sonicator behaves differently—probe-based instruments deliver powerful localized energy, while water-bath systems offer gentle, reproducible shearing. Parameters such as pulse time, duty cycle, and total duration must be empirically optimized for each sample type and volume.

For more fragile samples—or where epitope preservation is paramount—enzymatic digestion offers an alternative. Micrococcal nuclease (MNase) selectively digests accessible DNA between nucleosomes, enriching for mononucleosome-sized fragments. This approach, however, requires precise control over digestion time and enzyme concentration. Over-digestion can erase nucleosome positioning data, while under-digestion leads to incomplete fragmentation and poor enrichment. Some hybrid protocols combine mild sonication with limited MNase digestion to harness the advantages of both.

The troubleshooting phase of ChIP sample preparation is where technical knowledge meets practical instinct. A low chromatin yield may stem from incomplete lysis, ineffective crosslinking, or excessive sample dilution. Fragmentation too large suggests under-sonication or inadequate detergent concentration, while smearing or degradation points to insufficient protease inhibition or excessive heat during processing. Anticipating these issues—and recognizing their molecular fingerprints—can save valuable time and material.

Special care must also be taken when working with samples that yield naturally low chromatin quantities, such as sorted populations or rare cell types. In these cases, the temptation to scale up crosslinking reagents or sonication cycles must be resisted. Instead, consider concentrating the lysate, reducing buffer volumes, and maximizing antibody efficiency in subsequent steps. When preparing samples for histone mark ChIP, avoid any unnecessary freezing or repeated thaw cycles, which can distort histone modification patterns or fragment chromatin.

At the heart of this process is consistency. Sample preparation must be reproducible not only across replicates but across experiments, labs, and conditions. To that end, developing and adhering to a detailed checklist is invaluable. Ensure all reagents are freshly prepared. Confirm crosslinking times and temperatures are logged precisely. Always process samples on ice and include complete inhibitor cocktails in every buffer. Monitor DNA fragmentation with agarose gel electrophoresis or Bioanalyzer profiles. And above all, treat your chromatin with reverence—because every successful ChIP experiment begins long before the antibody binds.

In this silent but vital phase of ChIP, where cells surrender their secrets and chromatin is carved into insight-ready fragments, there is no room for haste or guesswork. The quality of sample preparation will echo through every downstream step—through antibody pulldown, through sequencing, and into the interpretation of regulatory landscapes. The genome may be the same in every cell, but how you choose to capture it makes all the difference.

8. CHROMATIN FRAGMENTATION

At the heart of every successful chromatin immunoprecipitation experiment lies a single pivotal transformation—the controlled fragmentation of chromatin. Until this point, cells have been harvested, crosslinked, and lysed. Proteins have been preserved in situ on their DNA binding sites, poised for interrogation. Yet without properly fragmented chromatin, even the most precise antibody will struggle to find its target. It is in this critical step—the breaking of chromatin into manageable, interpretable pieces—that the fate of a ChIP experiment is sealed. Fragment too much, and epitopes vanish. Fragment too little, and resolution collapses into unreadable noise.

The challenge is deceptively simple: reduce chromatin into 200–500 base pair fragments while preserving the protein-DNA interactions locked in by crosslinking. This is the window—the sweet spot—where most ChIP-grade antibodies operate best, and where the downstream sequencing resolution is highest. Achieving this requires finesse, not brute force. Every sample, every cell type, and every experimental design may demand a slightly different strategy. The two major paths forward are sonication and enzymatic digestion, each with its own philosophy of fragmentation.

Sonication, the mechanical shearing of chromatin through ultrasonic energy, is by far the most commonly used approach in crosslinked ChIP protocols. It is direct, fast, and scalable. The basic principle involves delivering high-frequency sound waves through a probe or bath, creating microscopic cavitation bubbles that collapse violently, shearing DNA and chromatin in the process. The duration, power, and duty cycle of this energy delivery determine the extent of fragmentation. But sonication is far from mindless pulsing—it is a precise and often temperamental dance.

Each sonicator is its own beast. Probe-based instruments offer the most intense energy delivery and can process samples rapidly, but they also generate heat and risk denaturing proteins or causing sample foaming. Bath-based sonicators deliver gentler, more uniform energy and are better for parallel processing, but they often require longer times to achieve optimal fragmentation. Researchers often begin by empirically determining ideal conditions: testing a range of cycles, usually 10 to 30 seconds per pulse, separated by equal rest intervals, and extending across multiple rounds of sonication. Over time, a pattern emerges—one where the chromatin falls consistently within the desired 200–500 bp window, and proteins remain structurally intact and immunoreactive.

Over-sonication is a common and silent killer in ChIP experiments. Excessive fragmentation can break apart nucleosomes, shear off DNA-binding domains, or even melt tertiary protein structures. The resulting chromatin may appear ideal on an agarose gel—clean, sharp bands within the right range—but immunoprecipitation fails because the epitopes have been lost. Conversely, under-sonication leaves behind long fragments—kilobases in length—that reduce the resolution of ChIP-seq, create ambiguous peaks, or result in low recovery. Worse still, these large fragments often carry multiple binding sites, leading to false associations and misinterpretation of protein localization.

Temperature is a hidden variable in sonication efficiency. Heat, generated by continuous energy pulses, can rapidly denature chromatin and antibodies alike. Keeping samples on ice, using water-cooled jackets, or inserting rest intervals between pulses becomes essential to maintain protein structure and prevent evaporation. Many sonicators now include temperature monitoring features, alerting researchers when internal limits are exceeded. It is a silent battle—between energy and preservation—and only those who listen closely to their samples will win.

For researchers who seek more elegance—or who are working with native chromatin not crosslinked by formaldehyde—enzymatic digestion offers an alternative. Micrococcal nuclease (MNase), a calcium-dependent endo-exonuclease, cuts DNA preferentially between nucleosomes, cleaving the linker regions and leaving behind nucleosome-sized fragments. In ideal conditions, this yields a laddered pattern of mononucleosomes (~147 bp), dinucleosomes (~300 bp), and higher-order units, clearly visible on an agarose gel. For experiments targeting histone modifications or nucleosome positioning, this enzymatic precision is unmatched.

Unlike sonication, which does not discriminate between accessible and protected chromatin, MNase digestion reflects the natural chromatin landscape. Regions of open chromatin are digested faster, while tightly packed heterochromatin remains protected. This property can be both a blessing and a curse. When correctly interpreted, MNase profiles provide insight into chromatin accessibility. But in ChIP experiments where uniform fragmentation is desired, this bias must be carefully controlled. Digestion conditions—enzyme concentration, time, and temperature—must be optimized meticulously, usually through titration curves and timed reactions.

The most delicate MNase digestions involve a balance between partial and complete digestion. Over-digestion collapses all chromatin into mononucleosomes, obliterating information about higher-order structure. Under-digestion results in large fragments that resist efficient immunoprecipitation. To walk this tightrope, researchers often perform time-course experiments, removing aliquots at various time points and assessing the DNA profile. The final decision point is not theoretical—it is visual. A 1.5% agarose gel or a capillary electrophoresis trace tells the story in bands and peaks. Are the fragments predominantly in the 150–300 bp range? Are higher-order nucleosomes still visible? The answers lie in these images.

Sample type plays a decisive role in fragmentation strategy. Cultured cells with soft membranes and loose chromatin are more amenable to both sonication and MNase digestion. But primary tissues—especially brain, muscle, or fibrotic samples—are challenging. Their dense extracellular matrices and rigid nuclei resist both mechanical and enzymatic access. In such cases, pre-treatment with dounce homogenization, collagenase, or detergent cocktails may be necessary.

Tissue-specific optimization is often the key to success, and there is no universal protocol that covers all sample types.

Fixation method also exerts a hidden influence on fragmentation efficiency. Crosslinking with formaldehyde stiffens chromatin and protects protein-DNA complexes, but it also reduces accessibility. Over-fixed samples resist shearing and may require higher sonication energy or longer MNase digestion. Under-fixed samples, while easier to fragment, may lose fragile interactions. Some protocols now advocate for sequential crosslinking with additional reagents like DSG or EGS, followed by optimized fragmentation routines. These hybrid strategies aim to balance preservation with accessibility, allowing researchers to fine-tune both the chemistry and physics of chromatin preparation.

Quality control is not optional—it is the insurance policy of every ChIP experiment. After fragmentation, a small aliquot of chromatin is always taken and de-crosslinked for fragment analysis. The gold standard remains gel electrophoresis: loading the DNA on a 1.5% agarose gel, staining with ethidium bromide or SYBR Safe, and visualizing under UV light. Ideally, the DNA smears cleanly between 200 and 500 bp, with minimal presence of high-molecular-weight bands. For finer resolution, instruments like the Agilent Bioanalyzer or TapeStation provide digital traces of fragment distribution, revealing sharp peaks, average sizes, and total yield. These traces are more than technical checks—they are diagnostic tools, guiding decisions about repeating fragmentation, proceeding to ChIP, or adjusting protocols.

Still, deciding between sonication and enzymatic digestion is not always straightforward. Researchers must weigh sample type, fixation status, target protein class, and experimental goals. Sonication is preferred for transcription factors, chromatin remodelers, and crosslinked material, offering uniform fragmentation across the genome. MNase excels in native ChIP, histone modification studies, and nucleosome mapping. For projects requiring both approaches—such as comparing native and crosslinked conditions—parallel experiments may be necessary.

To aid this decision, many protocols now include a fragmentation decision tree. It begins with a simple question: Is the chromatin crosslinked? If yes, proceed with sonication. If no, and native chromatin is available, test MNase digestion. For low-input samples or rare cells, choose the method that preserves epitopes best. If antibodies are sensitive to structural changes, avoid aggressive sonication. If sequencing resolution is paramount, optimize for narrow fragment size. And always validate with input controls—because even the most perfect fragmentation means nothing if the enrichment fails.

There is a quiet poetry to chromatin fragmentation. It is the moment when the vast architecture of the genome is broken down—not into chaos, but into signal-bearing fragments. Each sheared or digested piece carries a clue: a piece of DNA once touched by a protein, a whisper of transcription, a memory of repression or activation. These fragments, scattered across the genome, become the raw material for ChIP-seq peaks and epigenomic maps. They speak a language of structure and function, and fragmentation is the first stroke on their canvas.

In the end, whether you choose the thunder of sonication or the scalpel of enzymatic digestion, the goal remains the same: resolution. Precision. Clarity. Fragmentation is the bridge between biology and data, between cells and stories. Done well, it becomes invisible, a silent enabler of every discovery that follows. Done poorly, it shrouds the truth in noise and leads even the best antibodies astray. To fragment chromatin is not simply to break—it is to begin.

9. SELECTING AND VALIDATING ANTIBODIES

In the delicate architecture of a ChIP experiment, few components carry as much weight as the antibody. This molecule—elegant in its specificity and unforgiving in its flaws—can make or break the outcome of even the most meticulously planned study. Selecting and validating the right antibody is not merely a technical checkpoint; it is a strategic decision that demands a nuanced understanding of protein structure, epitope accessibility, immunogenic fidelity, and reproducibility. It is here, at the interface of biology and precision craftsmanship, that the experimental success of chromatin immunoprecipitation is forged.

The concept of a "ChIP-grade" antibody is often taken at face value in catalogs and online databases, yet it remains a term that carries variable meaning. A true ChIP-grade antibody must possess a trifecta of qualities: high specificity for the target protein or histone modification, strong and consistent affinity across different lots, and functionality in a chromatin context—where the protein is crosslinked to DNA and embedded in complex chromatin structures. This last criterion often differentiates a good antibody from a great one. An antibody that performs admirably in a Western blot may utterly fail in a ChIP assay because the epitope is masked or structurally constrained when crosslinked to DNA. Therefore, validation must go beyond superficial branding or hearsay—it must be rigorous, empirical, and purpose-built.

Commercial sources of antibodies are plentiful, ranging from long-standing biotechnology companies to niche suppliers specializing in epigenetics. Many vendors now include independent user reviews, application-specific citations, and validation data. Yet, the prudent researcher approaches such information with a blend of optimism and skepticism. Even well-reviewed antibodies can show lot-to-lot variability. The nature of polyclonal antibodies, in particular, introduces biological variability because they are derived from multiple B-cell clones, each targeting different epitopes. For this reason, many researchers prefer monoclonal antibodies, which offer greater specificity and batch consistency. However, the trade-off may lie in reduced flexibility or recognition of only a single, possibly masked epitope.

To truly validate an antibody for ChIP, experimental evidence must speak. One of the first steps is testing specificity through a dot blot or Western blot, ensuring the antibody binds the correct molecular weight protein and not others. This provides an initial readout of reactivity but is insufficient for ChIP purposes. More stringent validation requires chromatin context—using techniques like ChIP-qPCR at well-established positive and negative genomic loci. For histone modifications, these sites are often known from literature or public databases like ENCODE. For transcription factors, the approach may involve overexpression, knockdown, or ideally knockout models, where the absence of signal in the mutant background confirms specificity.

Controls are not optional; they are the compass by which one navigates the ambiguity of ChIP signal. A non-specific IgG control should always accompany the test antibody to gauge background binding. Similarly, a positive control antibody—such as anti-H3K4me3, known to bind active promoter regions—is invaluable for confirming the success of the ChIP protocol itself. These controls also help interpret whether weak enrichment is due to antibody failure or procedural inefficiency. Species compatibility must be heeded as well—antibodies raised in rabbit may not be ideal when used with secondary anti-mouse detection systems, or vice versa. The small details, from host species to buffer compatibility, can influence the final data.

For labs working with tagged proteins—such as FLAG, HA, or MYC tags—the ChIP process can be simplified using tag-specific antibodies. These systems bypass the variability of endogenous protein recognition and are particularly useful in model systems with stable transgenic expression. Still, even tagged systems require validation to ensure that tag placement does not hinder protein function or DNA binding, and that the tag antibody pulls down chromatin in a manner that recapitulates native patterns.

Batch testing is the unsung hero of long-term ChIP projects. An antibody that performed flawlessly last year may fail in its new lot. Performing pilot ChIP experiments with each new batch—preferably on a small scale using known target loci—ensures continuity and reliability. This becomes especially important in multi-year projects, clinical sample studies, or high-throughput ChIP-seq pipelines where reproducibility is non-negotiable.

Troubleshooting is often where the antibody's true colors are revealed. Excessive background may suggest poor specificity, improper blocking, or faulty washing steps. Weak or no signal may arise from insufficient affinity, degraded antibody, or poor chromatin quality. In such cases, titrating antibody amounts, adjusting incubation times, or switching buffer formulations may help. But sometimes, the solution lies in returning to first principles: reevaluating the antibody altogether.

Toward the end of this critical selection process, a checklist begins to form—one not of protocol steps, but of intellectual diligence. Is the antibody validated in ChIP by independent sources? Is it monoclonal or polyclonal? Are there known citations for its use? Are the host species and epitope details compatible with your experimental design? Has a pilot study confirmed specificity in your own system? Can it be trusted not just in a publication, but across replicates, platforms, and time?

As a researcher standing at the threshold of ChIP experimentation, choosing an antibody is akin to selecting a lens for a telescope—you may be staring at the vast chromatin universe, but clarity depends entirely on the optic through which you view it. The right antibody transforms chaotic noise into meaningful signal. It bridges invisible protein-DNA interactions into interpretable peaks. And ultimately, it carries your experiment from question to discovery—anchored by the simplest yet most powerful element in your toolkit: trust in the molecule that binds.

10. IMMUNOPRECIPITATION

In the intricate orchestration of a ChIP experiment, immunoprecipitation stands as its pivotal act—the moment when the invisible architecture of protein-DNA interactions is seized from solution and made tangible. It is here that theory meets touch, where antibodies reach into the chaotic mix of chromatin fragments and selectively retrieve the complexes we seek to understand. This step, deceptively simple in concept, is laden with nuance. And mastering it is akin to learning the fine motor skills of molecular sleuthing—where every condition, every reagent, and every pause in the protocol has a consequence.

The goal of immunoprecipitation is elegantly straightforward: to capture, enrich, and isolate DNA-bound proteins of interest using specific antibodies. But achieving this with precision and minimal background noise requires a marriage of biochemical insight and empirical rigor. The choice of support matrix—whether magnetic beads or agarose slurry—is often the first fork in the road. Magnetic beads, particularly those coated with Protein A or Protein G, offer convenience, rapid handling, and clean separation using magnets rather than centrifugation. Agarose beads, on the other hand, have long been the standard in classic immunoprecipitation protocols, often preferred for certain binding efficiencies or when cost is a constraint. Protein A and Protein G differ in their affinities for immunoglobulin subtypes; the optimal choice depends on the species and subclass of the antibody used. Rabbit IgG tends to bind strongly to Protein A, whereas mouse monoclonals often favor Protein G. Some vendors now offer beads conjugated with both proteins or engineered chimeras to bridge this gap.

Before antibody and chromatin even meet, the beads are pre-treated—blocked, washed, and pre-cleared—to reduce non-specific binding. This often involves incubating the beads in a blocking buffer containing bovine serum albumin (BSA), salmon sperm DNA, or non-fat dry milk. These blockers saturate sticky sites on the beads that might otherwise attract DNA, histones, or other cellular debris. A well-blocked bead is the molecular equivalent of a clean fishing hook: invisible to everything except the intended catch.

Then comes the heart of the process—incubating the sheared chromatin with the antibody. This is not a brute-force reaction; it is a slow, patient dance of diffusion, recognition, and binding. Typically, incubation is done at 4°C for 4 to 16 hours with gentle rotation. The chromatin must remain intact and stable, the antibody fully active, and the temperature low enough to prevent degradation. The salt concentration of the binding buffer plays a crucial role here. Too low, and non-specific interactions flourish. Too high, and even specific complexes may fall apart. Most protocols strike a balance around 150 mM NaCl, with additives like EDTA and NP-40 to prevent aggregation or proteolysis.

Once the antibody-chromatin complex is formed, it is captured by the beads—either pre-coupled or added afterward—and subjected to a series of washes that are as critical as the binding step itself. These washes, typically performed at 4°C, serve as a molecular filtration system. Low-stringency buffers remove loosely associated proteins, while high-stringency buffers—containing elevated salt concentrations or detergents—strip away all but the most tightly bound complexes. The number and composition of these washes must be tuned to the sensitivity of the assay and the nature of the target. Histone modifications, being tightly bound and abundant, can withstand harsher washes. Transcription factors, often present in low copy number and with weaker binding kinetics, may be lost if the stringency is too high. It's a balancing act, one honed by trial, experience, and sometimes sheer instinct.

Immunoprecipitation protocols often diverge here, depending on whether the target is a histone mark or a DNA-binding factor. Histone ChIP typically benefits from a more straightforward protocol, with less reliance on extensive optimization. The epitopes are exposed and abundant, the interactions stable. Transcription factor ChIP, by contrast, is a game of margins. Success requires optimized crosslinking, higher cell input, longer incubation times, and sometimes even dual crosslinking strategies (e.g., formaldehyde plus DSG or EGS) to stabilize protein complexes.

Elution is the final act of retrieval—the unhooking of the target complex from the beads. Several strategies are employed, each with its merits. Heat elution, often at 65°C with SDS-containing buffers, denatures proteins and breaks antibody-antigen interactions, releasing the chromatin. This method is straightforward and effective for most targets. Low pH elution, using buffers like glycine-HCl at pH 2.5–3.0, offers a gentler alternative that preserves protein integrity for downstream analysis, though it requires immediate neutralization. Peptide competition—using an excess of the epitope peptide to outcompete the antigen—offers specificity, but is only applicable when synthetic peptides are available and affordable.

In more advanced applications, researchers turn to sequential ChIP, or Re-ChIP, wherein two immunoprecipitations are performed in series. This method allows interrogation of co-occupancy—whether two proteins or modifications exist on the same DNA molecule. It is powerful, but technically demanding, requiring careful optimization of elution conditions to preserve native chromatin structure between rounds.

The scale of immunoprecipitation is another consideration. ChIP protocols are notoriously hungry for input material. Scaling up from 1 million to 10 million cells may be necessary for rare targets or weakly binding antibodies. Yet even at small scale, precision matters. Antibody concentration, bead volume, and incubation kinetics must all be proportionally adjusted to avoid saturation or waste.

To assist in navigating these variables, a comparison chart of immunoprecipitation protocols can be invaluable. It summarizes recommended conditions for histone marks (e.g., H3K27ac, H3K9me3) versus transcription factors (e.g., CTCF, p53), highlights suggested wash stringencies, elution methods, and incubation durations, and even notes common pitfalls. Such a chart is not a rigid script but a roadmap—a place to begin before making the method your own.

In the end, immunoprecipitation is less a step than an art form. It demands patience, preparation, and the humility to adjust. It is where chemistry and biology meet, held together by faith in the antibody, the protein, and the silent code of chromatin. Done well, it yields the fragments that map our genomes' regulatory logic. Done poorly, it yields noise dressed as data. But even in failure, it teaches. And in its mastery, it opens the gates to discovery.

11. REVERSE CROSSLINKING AND DNA PURIFICATION

Once the immunoprecipitation step is complete and the beads have yielded their precious cargo of chromatin complexes, the ChIP experiment enters a critical transitional phase—transforming bound protein-DNA interactions into purified DNA suitable for downstream analysis. It is the moment when the ephemeral contacts that preserved cellular memory are gently dismantled, revealing the bare genomic sequences beneath. This chapter takes readers into that transformative process: the reverse crosslinking and purification phase that releases the genetic script from the grasp of regulatory proteins and histones.

Crosslinking, typically initiated with formaldehyde, serves as the molecular fixative that stabilizes protein-DNA interactions in vivo. However, to analyze the DNA, those links must be carefully broken. This is not a step to rush. The reversal must be thorough but gentle, precise but unyielding—leaving no residue of protein, crosslinker, or RNA behind. The conditions under which this happens can determine not only the quantity of DNA recovered but also its usability in sensitive techniques like qPCR, sequencing, or library construction.

The most widely employed method of reversing formaldehyde crosslinks is thermal treatment. Heating the immunoprecipitated samples, typically at 65°C for 4 to 16 hours, efficiently breaks the methylene bridges that covalently link proteins to DNA. This is usually done in the presence of elution buffer containing detergents such as SDS, which help solubilize protein complexes and denature histones or transcription factors. Longer incubations ensure more complete reversal, but they also introduce risks: evaporation, DNA degradation, or sample loss if not carefully monitored. Proper sealing of reaction tubes and maintaining a humidified incubator or using oil overlays for small-volume reactions are common preventative measures.

Chemical methods also exist for de-crosslinking, particularly for dual crosslinking strategies involving reagents like disuccinimidyl glutarate (DSG) or ethylene glycol bis(succinimidyl succinate) (EGS), which target protein-protein interactions. These require more specialized buffers or reducing conditions to reverse. However, even in these cases, the final thermal step at 65°C remains a universal denouement—ensuring formaldehyde-mediated bonds are fully broken.

After thermal de-crosslinking, the next step is enzymatic cleanup. RNA, while not the primary target of ChIP, often co-purifies with chromatin and can interfere with downstream quantification or library construction. RNase A treatment, typically at 37°C for 30 minutes, efficiently degrades contaminating RNA. Following that, Proteinase K is introduced—an enzyme that digests denatured proteins, histone cores, and residual antibodies that may still cling to the DNA. This step, usually performed at 55°C for an hour or more, ensures that what remains is pure nucleic acid, free from enzymatic inhibitors or spectrophotometric noise.

DNA purification is the final gatekeeper before analysis. Two major approaches dominate this landscape: phenol-chloroform extraction and silica spin column purification. Each comes with trade-offs.

Phenol-chloroform extraction is a classical organic method, revered for its completeness but feared for its harshness. The process involves mixing the aqueous sample with equal volumes of phenol and chloroform (sometimes in a 1:1 or 25:24:1 ratio with isoamyl alcohol), centrifuging to separate phases, and then recovering the upper aqueous phase containing DNA. While highly effective at removing proteins and lipids, phenol is toxic and volatile, requiring fume hood conditions and careful handling. Additionally, DNA must be precipitated—usually with ethanol or isopropanol in the presence of salt and a carrier like glycogen or linear acrylamide—before it can be resuspended in buffer.

Spin columns, by contrast, offer speed, safety, and reproducibility. Commercial kits use silica membranes that selectively bind DNA in high-salt conditions. After binding, contaminants are washed away with ethanol-based buffers, and the DNA is eluted in water or Tris-EDTA. The efficiency of this method depends on DNA size, salt concentration, and column capacity, but for most ChIP applications—especially when working with small inputs—spin columns are the preferred choice. Some kits are optimized for low input samples, down to picogram levels, and offer improved recovery rates with minimal binding bias.

Troubleshooting DNA recovery is often necessary. Low DNA yield may stem from inefficient crosslink reversal, incomplete protein digestion, DNA shearing, or poor binding to the column. Residual protein contamination can obscure quantification or inhibit enzymatic steps downstream. The key to resolution often lies in incremental optimization—extending incubation times, verifying buffer pH, or introducing additional washes.

The way DNA is handled at this stage matters enormously. Vortexing, excessive pipetting, or repeated freeze-thaws can shear DNA and reduce yield. Low-retention tubes, gentle mixing, and minimal transfers are recommended. When recovering small amounts of DNA, even the choice of elution volume (usually 10–50 μL) can determine concentration and downstream performance. Some protocols suggest heating the elution buffer to 60°C and letting it incubate on the column for a few minutes to enhance yield.

Once DNA is purified, its quality must be assessed. Quantification using fluorometric assays like Qubit is highly recommended, as these dyes selectively bind double-stranded DNA and exclude RNA or free nucleotides. Spectrophotometric methods like Nanodrop are faster but less accurate at low concentrations and more sensitive to contamination. The A260/280 ratio gives an estimate of protein contamination, with values around 1.8 considered pure for DNA. A260/230 ratios, ideally above 1.5, indicate minimal organic or salt carryover.

Small input protocols deserve special mention. In low-cell ChIP or single-cell adaptations, even femtogram-scale DNA must be recovered reliably. This requires modified column protocols, magnetic bead-based capture methods, or carrier nucleic acids. Kits tailored for these applications often reduce wash volumes, avoid ethanol drying steps, and provide enhanced elution buffers to maximize retention.

Ultimately, the aim is to obtain DNA that is clean, intact, and truly representative of the initial chromatin landscape. At this juncture, the original spatial interactions between proteins and the genome have been resolved into a sequenceable form. The success of everything that follows—qPCR, library preparation, sequencing, and bioinformatics—depends on this phase being executed with precision.

Reverse crosslinking and purification may appear to be a winding down of the ChIP protocol, but in reality, it is its crescendo. It is where fidelity is preserved or lost, where a clean window into epigenetic regulation is either opened or fogged by artifacts. When performed with care, this phase delivers DNA that not only amplifies cleanly and maps accurately but reflects the living reality of transcriptional regulation in cells. It is, in essence, the gentle unwrapping of chromatin's most guarded secrets—one molecule at a time.

12. QUALITY CONTROL IN ChIP EXPERIMENTS

Every successful chromatin immunoprecipitation experiment tells a story—one of molecular choreography between proteins and DNA, one that reveals how the genome is not just a static blueprint but a dynamic, responsive system. But even the most elegant ChIP story can become a tale of misdirection if its plot is marred by poor quality control. Quality control (QC) in ChIP is not an afterthought; it is the scaffold upon which scientific reliability stands. It is what separates an artifact from evidence, noise from narrative, and illusion from insight. In this chapter, we venture into the nuanced and essential world of ChIP quality control—the vigilant checkpoints that verify whether each step of the experiment has fulfilled its promise.

From the moment cells are harvested to the final moment DNA is ready for sequencing or amplification, every phase offers opportunities for success or failure. It begins with fragmentation—whether by sonication or enzymatic digestion—where the first QC checkpoint arises. The size of fragmented chromatin is critical. If too large, the resolution of DNA-protein mapping suffers; if too small, valuable epitope structures or DNA integrity can be lost. Here, gel electrophoresis becomes the first gatekeeper. Running an aliquot of the sheared chromatin on a 1–2% agarose gel allows visualization of the DNA smear. Ideally, this smear should range between 200 and 500 base pairs, a size optimal for both ChIP enrichment and sequencing compatibility. The Bioanalyzer or TapeStation further refines this assessment, offering digital electropherograms with precise fragment distribution profiles, allowing for sensitive comparison across samples, batches, or replicates.

But fragmentation is only the beginning. Once immunoprecipitation is complete, the next layer of QC focuses on enrichment—determining whether the antibody actually pulled down the target protein-DNA complex with specificity and efficiency. This is where ChIP-qPCR emerges as a powerful sentinel. By using primers that flank well-characterized genomic regions—typically known positive and negative binding sites for the protein of interest—researchers can measure fold enrichment and background signal. For histone modifications, these target regions are often derived from published ChIP-seq datasets or public epigenomic consortia like ENCODE. For transcription factors, validated enhancers, promoters, or known motif-containing loci are used.

The importance of primer design cannot be overstated. Good ChIP-qPCR primers should amplify 80–150 bp products and reside in genomic areas free of repetitive sequences or structural anomalies. Positive control primers target regions where the protein is known to bind; negative control primers target genomic deserts or regions where the factor is not expected to be present. The difference in Ct values—converted into fold enrichment—serves as a quantitative litmus test for success. Typically, a fold enrichment greater than 5–10× over the IgG control, and with minimal amplification in negative regions, signifies a robust experiment.

One of the emerging pillars of ChIP quality control is the use of spike-in controls. These are exogenous chromatin fragments—often from Drosophila or yeast—added in equal amounts to each sample before immunoprecipitation. They come paired with antibodies specific to a histone modification or DNA-binding protein from the same species. Since these spike-ins are invariant across conditions, they offer a normalization scaffold for technical variability, antibody efficiency, or sample input inconsistencies. This is particularly crucial in ChIP-seq studies where batch effects can skew global signal distribution and lead to erroneous biological interpretations. Normalizing against spike-ins helps decouple experimental signal from background drift.

Cross-reactivity and noise—two ghosts that haunt ChIP assays—also require vigilance. Cross-reactivity may stem from the antibody binding off-target proteins or epitopes, resulting in false enrichment. Noise, on the other hand, often emerges from over-crosslinked chromatin, poor antibody specificity, or insufficient washing. These problems manifest in elevated background signal in negative control qPCR reactions or smeared bands in gels. Background noise can also inflate apparent enrichment in non-target regions, making biological interpretation murky. The solution lies in layer-by-layer validation: running IgG controls, comparing multiple antibodies for the same target, and always benchmarking against known standards.

Batch-to-batch consistency is another cornerstone of ChIP QC. Whether dealing with different lots of antibody, changes in cell culture conditions, or new personnel performing the assay, ChIP is particularly susceptible to subtle shifts that amplify downstream. Running biological replicates—independent samples processed in parallel—allows for the estimation of biological variability and supports statistical confidence in observed patterns. Technical replicates—splits of the same sample processed identically—help evaluate procedural reproducibility. High correlation between biological replicates, whether measured by qPCR or genome-wide sequencing, is a hallmark of well-controlled ChIP studies.

A critical aspect of QC is knowing when and how to troubleshoot. Low enrichment might point to weak antibody binding, insufficient chromatin, or degraded DNA. Excessive background could indicate over-crosslinking, antibody cross-reactivity, or contamination in reagents. Poor DNA recovery may result from inadequate elution or incomplete reversal of crosslinking. Each of these signs points to a stage in the protocol where refinement is needed. Keeping a detailed experimental log, including reagent lot numbers, incubation times, temperatures, and instrument settings, provides a roadmap for troubleshooting and iteration.

Ultimately, quality control is not a discrete step—it is an attitude, a discipline that permeates every part of the ChIP workflow. It teaches researchers to think like detectives, to question every band, every amplification curve, and every unexpected peak. It empowers them to challenge assumptions, revisit protocols, and refine techniques. Most importantly, it builds confidence—not only in the data but in the conclusions drawn from them.

To perform ChIP without QC is to peer through a fogged lens; the shapes may be intriguing, but their outlines cannot be trusted. With rigorous quality control, however, each fragment of chromatin recovered becomes a piece of a reliable puzzle—a glimpse into the regulatory lexicon of the cell. The data gained from a well-controlled ChIP experiment do more than show where a protein binds; they illuminate how the genome is interpreted, rewritten, and remembered. Quality control makes that interpretation trustworthy—and the story of the genome, all the more clear.

A final layer of quality control lies in data management and documentation. Beyond wet-lab precision, reproducibility is also rooted in transparent record-keeping. Researchers are encouraged to document antibody catalog numbers and lot identifiers, chromatin input amounts, sonication cycles or enzyme digestion conditions, buffer recipes, and detailed timelines for each step. This level of documentation not only facilitates internal consistency but also supports compliance with publication standards and peer review scrutiny. In the era of FAIR data principles—Findability, Accessibility, Interoperability, and Reusability—such meticulousness is no longer optional. It is an ethical imperative.

For those transitioning into genome-wide ChIP-seq workflows, QC takes on an even broader computational dimension. Tools like FastQC, Picard, and deepTools offer statistical insights into library complexity, read quality, GC bias, duplication rates, and mapping efficiency. Enrichment over input control, peak-calling stringency, and the fraction of reads in peaks (FRiP) score provide further metrics to distinguish signal from background. Importantly, these tools do not replace experimental QC; rather, they enhance it—validating that what was enriched in the tube translates into meaningful patterns in the genome browser. The intersection of experimental rigor and computational integrity is where modern ChIP quality control reaches its zenith.

Training and calibration are also worth mentioning. In labs with multiple users or collaborators, standardization becomes a challenge. Implementing internal quality metrics—such as requiring specific fold-enrichment thresholds or DNA purity ratios—can align efforts and expectations. Running periodic test samples across hands, reagents, or machines helps calibrate variability. For clinical or industrial ChIP applications, this standardization may take the form of Good Laboratory Practice (GLP) protocols or ISO certifications, with formal quality audits and performance benchmarks.

To support this holistic approach, many labs adopt a ChIP QC dashboard—a centralized tracking document or digital tool that logs sample details, QC outputs, and batch metrics. Whether as a spreadsheet, a lab notebook insert, or a custom-built app, the dashboard serves as a living memory of experimental fidelity. It enables researchers to spot trends, flag recurring issues, and compare new data against a baseline of expected performance.

As the field of chromatin research grows more nuanced, the expectations for data reliability increase. No longer is it enough to detect a binding peak; one must also prove its validity, reproducibility, and biological relevance. This chapter empowers researchers not just with protocols, but with a mindset—an unwavering commitment to precision, transparency, and vigilance.

In summary, quality control is the heartbeat of every successful ChIP experiment. From chromatin fragmentation and antibody performance to DNA purity and data interpretation, QC ensures that each link in the experimental chain holds firm. By embracing QC as an integrated, continuous process—not a hurdle, but a habit—researchers gain more than clean data. They gain trust in their discoveries, respect in their community, and a lasting contribution to the genomic map of life.

And in that careful, persistent pursuit of quality, the chromatin landscape begins to speak with clarity—revealing not just where proteins bind, but how they orchestrate the symphony of cellular identity and fate.

13. ChIP-qPCR TARGETING

Among the most accessible and widely adopted approaches in chromatin immunoprecipitation workflows, ChIP-qPCR occupies a unique position at the intersection of specificity, affordability, and interpretability. While high-throughput sequencing technologies like ChIP-seq have transformed the landscape of epigenomic research, the simplicity and efficiency of ChIP-qPCR make it indispensable for focused studies, validation experiments, and laboratories operating under resource constraints. At its core, ChIP-qPCR enables researchers to precisely quantify the enrichment of DNA sequences bound by a protein of interest at predefined genomic loci. With the right design, proper controls, and sound experimental planning, ChIP-qPCR becomes a powerful lens to examine transcription factor occupancy, histone modification landscapes, and chromatin state transitions across various biological conditions.

The strength of ChIP-qPCR lies in its ability to interrogate known promoter, enhancer, or intergenic regions with high sensitivity. To begin such investigations, thoughtful primer design becomes paramount. Primer pairs are tailored to amplify regions where the binding of a transcription factor or the presence of a specific histone mark is expected or has been predicted through prior studies or bioinformatic tools.

Promoter regions are often the first choice, especially when targeting transcription factors or activating histone marks like H3K4me3. Enhancer regions, more distal in nature, are typically targeted when investigating histone acetylation marks such as H3K27ac or factors like p300. Intergenic regions, often devoid of signal, serve as negative controls. To design primers that function optimally, the amplicon size is generally kept between 80 and 150 base pairs. This range ensures compatibility with fragmented chromatin derived from sonication or enzymatic digestion. Primers must also be tested for self-dimerization, secondary structures, and non-specific binding. Software like Primer3, NCBI Primer-BLAST, or commercial qPCR design tools can assist in creating robust primer sets that yield single, specific amplification products.

Once primers are validated, the next critical step is understanding how to quantitatively assess enrichment. The gold standard for ChIP-qPCR quantification is the comparative Ct method, commonly referred to as the 2^-ΔΔCt method. This approach provides a fold-change measure by normalizing the Ct value of the ChIP sample to an input control, and then comparing this normalized value to a reference condition or locus. The input sample, typically 1–10% of the chromatin taken before immunoprecipitation, accounts for differences in chromatin accessibility and overall DNA recovery. The ΔCt is calculated as the difference between the ChIP and input Ct values for a given target. The ΔΔCt then compares this value to a control region or experimental baseline, allowing researchers to interpret relative enrichment levels across multiple samples or treatments. Importantly, this method assumes comparable PCR efficiencies between target and control primers, which must be verified during assay setup.

Input normalization is just one part of ensuring quantitative accuracy. Control regions, both positive and negative, are essential for interpreting ChIP-qPCR results meaningfully. A positive control region is a genomic locus where the protein or modification is expected to be consistently present—for example, the GAPDH promoter for H3K4me3 or the MYC enhancer for H3K27ac. Negative control regions, such as intergenic deserts or promoters of genes not expressed in the cell type under study, should yield minimal signal. By comparing target enrichment to these controls, researchers can distinguish true biological signal from background noise or non-specific pulldown. Moreover, these controls are instrumental when troubleshooting unexpected results or low enrichment levels.

As with all PCR-based assays, technical artifacts can distort results if not carefully mitigated. Primer-dimer formation, secondary structures, and inefficient amplification can undermine the fidelity of quantification. Melt curve analysis and agarose gel electrophoresis are simple but effective methods to verify the specificity of amplification products. Reagents such as SYBR Green or probe-based detection systems like TaqMan may be selected depending on the desired level of sensitivity and specificity. Each qPCR reaction should ideally be run in technical triplicates, and biological replicates must be included to capture natural variability. Efficiency of amplification should be determined by running standard curves with serial dilutions of input DNA, ensuring that the reaction falls within the acceptable efficiency range of 90%–110%.

Case studies underscore the practical applications of ChIP-qPCR across diverse research contexts. For instance, in a study investigating the regulation of a metabolic gene under hypoxic conditions, ChIP-qPCR revealed increased binding of HIF-1α at the gene's promoter after just four hours of low oxygen exposure. Another example involved histone modification profiling in neural stem cells: researchers used ChIP-qPCR to show a sharp rise in H3K27me3 at differentiation-repressed loci as the cells transitioned from pluripotency. Such data would have been difficult to validate or monitor in real time with ChIP-seq alone, particularly due to the costs and timelines involved. In disease research, especially in cancer biology, ChIP-qPCR has become a cornerstone technique to investigate changes in chromatin state or transcription factor activity in patient-derived samples or disease models, where sample availability may be severely limited.

Interpreting ChIP-qPCR results involves more than just comparing Ct values or calculating fold enrichment. Statistical significance must also be determined to ensure that observed differences are not due to chance. Software such as GraphPad Prism, R, or specialized qPCR analysis tools can be used to perform Student's t-tests or ANOVA on replicate measurements. Error bars representing standard deviation or standard error of the mean are crucial for visualizing variability and conveying experimental robustness. Replicates—both biological and technical—are not optional but essential. Without them, any quantitative conclusions drawn from ChIP-qPCR remain speculative at best.

One of the most powerful applications of ChIP-qPCR is in validation of ChIP-seq data. After a genome-wide analysis identifies potential binding sites or enriched regions, a focused ChIP-qPCR assay can confirm whether these peaks are biologically relevant and reproducible. Conversely, qPCR can be used to pre-screen loci before committing to a full-scale ChIP-seq run, particularly in pilot experiments or under new experimental conditions. This dual utility makes ChIP-qPCR a linchpin method in any chromatin research pipeline.

For researchers just beginning their journey in epigenetics or chromatin biology, ChIP-qPCR offers a manageable entry point. It does not require extensive bioinformatics training, nor does it rely on expensive sequencing infrastructure. Yet, when properly executed, it delivers insights with high resolution and quantitative precision. The reduced cost and faster turnaround time compared to sequencing approaches allow for rapid hypothesis testing and protocol optimization. Moreover, its scalability means that even small laboratories can conduct meaningful chromatin studies, opening doors to broader participation in genomic science.

Ultimately, the success of ChIP-qPCR depends not on any single step, but on the harmonious integration of primer design, immunoprecipitation quality, control selection, and data analysis. Each element reinforces the others, creating a feedback loop of refinement and validation. In the hands of a thoughtful scientist, ChIP-qPCR becomes more than a technique—it becomes a language through which chromatin speaks. As molecular biology continues to evolve, this language will remain a vital tool for decoding the stories written in DNA, one target at a time.

Beyond its technical merits, ChIP-qPCR has also become a valuable teaching tool in academic and clinical laboratories. Because it requires a sequence-specific hypothesis, the method encourages researchers to think critically about gene regulation, chromatin accessibility, and transcriptional control. Students learning ChIP-qPCR must engage with questions such as: Which genomic regions are likely to be bound by this transcription factor? How might histone modifications influence gene expression in this cell type? What is an appropriate negative control? These foundational questions not only strengthen experimental design but also cultivate a deeper understanding of epigenetic mechanisms.

In clinical research, ChIP-qPCR is increasingly being adapted for diagnostic and prognostic purposes. Certain cancers, for instance, exhibit aberrant enrichment of activating or repressive histone marks at oncogenes and tumor suppressors. ChIP-qPCR allows precise measurement of these chromatin changes in biopsy samples, providing potential biomarkers for early detection or treatment monitoring. For example, in acute myeloid leukemia (AML), differential H3K27me3 occupancy at the *HOXA9* locus correlates with disease progression. A targeted ChIP-qPCR assay developed around this finding can offer clinicians real-time chromatin profiling as part of a personalized medicine strategy. Though still in early stages of clinical translation, such applications signal a future where chromatin state is part of routine diagnostic panels.

When integrated into time-course studies, ChIP-qPCR reveals the dynamics of protein-DNA interactions and chromatin modifications across biological processes. For instance, during cell cycle progression, researchers have used ChIP-qPCR to track histone H4 acetylation across key cell cycle gene promoters, observing oscillatory patterns that align with gene activation phases. Similarly, in response to drug treatments or environmental stressors, time-resolved ChIP-qPCR can illustrate transient chromatin remodeling events, shedding light on regulatory mechanisms that may be missed by static assays. These dynamic measurements, particularly when aligned with transcriptomic data, allow for the construction of regulatory models that tie chromatin changes to functional outcomes.

Normalization strategies continue to evolve in the ChIP-qPCR field, particularly when dealing with complex or heterogeneous tissues. One approach gaining traction is the use of spike-in chromatin from a different species (e.g., *Drosophila melanogaster*) as an internal control. This allows for more accurate cross-sample comparisons and can help mitigate variability in chromatin preparation or antibody pulldown efficiency. Though more complex to set up, such strategies enhance reproducibility and bring ChIP-qPCR closer to the rigor of sequencing-based methods.

Another frontier in ChIP-qPCR development lies in multiplexed detection. While conventional qPCR instruments typically analyze one target per well, emerging digital PCR platforms allow for simultaneous quantification of multiple genomic regions using fluorophore-tagged probes. This enables more comprehensive profiling within a single experiment and can be especially useful when investigating enhancer-promoter interactions or combinatorial histone modifications. Though currently more expensive and technically demanding, multiplexed ChIP-qPCR represents a promising direction for labs seeking to expand data yield without transitioning fully to next-generation sequencing.

Cross-laboratory reproducibility remains a crucial concern in ChIP-based assays, and ChIP-qPCR is no exception. To enhance standardization, several consortia and guidelines have emerged. The ENCODE and modENCODE projects have published best practices for antibody validation, input normalization, and qPCR reporting standards. Following these recommendations ensures that ChIP-qPCR data can be compared across experiments and even institutions, enabling meta-analyses and cross-study validation. Journals increasingly require these quality controls to be reported in methods sections, a shift that raises the overall rigor and transparency of ChIP-based research.

While much of the focus in ChIP-qPCR literature centers on transcription factors and histone modifications, the method can also be applied to non-traditional chromatin-associated proteins. For example, DNA methylation readers, chromatin remodelers, and long non-coding RNAs tethered to chromatin can all be profiled through modified ChIP-qPCR protocols. The versatility of the approach continues to grow, with innovations like ChIP-exo-qPCR and CUT&RUN-qPCR hybrid methods blending features from other technologies to enhance resolution and specificity. These refinements keep ChIP-qPCR relevant in an age of constant methodological evolution.

Importantly, the success of ChIP-qPCR depends on rigorous antibody validation. Antibodies must be highly specific to the target protein or histone mark, with minimal cross-reactivity. Poor antibody performance leads to misleading results, especially given the narrow window of detection that qPCR offers. Vendors now provide validation data against knockout samples, and many researchers perform preliminary Western blot or dot blot testing to confirm antibody functionality before proceeding to ChIP. Ideally, lot-to-lot consistency and proper storage conditions are maintained to ensure long-term reproducibility of results.

Even when all technical aspects are optimized, the ultimate value of ChIP-qPCR lies in how the data is interpreted and contextualized. Fold enrichment values, though informative, do not exist in isolation. They must be tied to changes in gene expression, phenotypic outcomes, or signaling pathway activity. Integrating ChIP-qPCR results with RNA-seq, ATAC-seq, or chromatin conformation assays such as Hi-C can create a multidimensional view of gene regulation. In this framework, ChIP-qPCR provides the ground truth—a fine-grained, quantitative snapshot that validates broader trends observed in genome-wide studies.

In conclusion, ChIP-qPCR stands as one of the most robust, versatile, and informative techniques for exploring the chromatin landscape. Its blend of affordability, accessibility, and quantitative precision makes it an essential tool in the molecular biologist's arsenal. Whether used to validate ChIP-seq data, investigate a novel transcription factor, or track dynamic epigenetic changes, ChIP-qPCR delivers insights that are both actionable and deeply revealing. For labs seeking high-impact results without the overhead of high-throughput platforms, ChIP-qPCR offers a compelling solution—a beacon of clarity in the intricate dance of DNA and its many regulators. And for scientists driven by curiosity, rigor, and a hunger to decode the genome's hidden scripts, ChIP-qPCR provides both the compass and the map.

14. ChIP-CHIP MAPPING

Before the era of high-throughput sequencing, before the genomes of cells could be read like open books, there was ChIP-chip—a pioneering marriage between immunoprecipitation and microarray technology that opened a new window into the regulatory landscape of DNA. Though now often viewed as a precursor to ChIP-seq, ChIP-chip once stood at the cutting edge of functional genomics. It brought with it the first glimpses of genome-wide protein-DNA interactions, allowing researchers to move beyond single loci and into broader terrains of transcriptional control. This chapter traces the method's foundations, practical execution, analytical frameworks, and historical significance, illuminating why this analog giant still holds relevance in specific scientific niches.

The process begins, as always, with a question: Where does a protein of interest bind within the genome? To answer this, chromatin is harvested and immunoprecipitated using a high-affinity antibody against the desired transcription factor, histone modification, or chromatin-associated protein. The immunoprecipitated DNA—now enriched for binding sites—is purified and labeled with fluorescent tags. A reference sample, often the input chromatin or a mock IP control, is labeled with a different fluorophore. These two DNA pools—experimental and control—are then co-hybridized onto a glass slide spotted with thousands to millions of oligonucleotide probes corresponding to genomic regions of interest. Fluorescence ratios at each probe site offer a snapshot of enrichment, flagging the places in the genome where the protein was most likely bound.

The power of ChIP-chip lies in its ability to simultaneously assess binding across large genomic stretches, particularly promoter regions, CpG islands, or other functionally annotated sequences. Early custom arrays focused on a few hundred gene promoters; later, commercial tiling arrays expanded to cover entire chromosomes or even whole genomes, depending on the organism and platform. This allowed researchers to survey the binding profiles of factors like p53, c-Myc, and RNA polymerase II, revealing not only canonical binding motifs but also novel targets and regulatory relationships. In epigenetic studies, ChIP-chip was instrumental in profiling patterns of histone acetylation and methylation across developmental stages, disease states, and drug responses, providing some of the first maps of chromatin landscapes in higher eukaryotes.

Yet, the technique is not without its limitations. At the heart of ChIP-chip is the microarray—a grid of probes fixed in space, each with a predefined sequence. This rigidity imposes constraints on genomic coverage and resolution. Binding sites located in repetitive elements, low-complexity regions, or poorly annotated loci may escape detection if not represented on the array. Similarly, resolution is capped by probe density and spacing; while modern arrays may feature probes every 50–100 base pairs, they cannot match the single-nucleotide resolution of sequencing. Signal bleed, cross-hybridization, and saturation effects can further confound quantitation, particularly when enrichment levels differ by orders of magnitude.

To mitigate these challenges, researchers developed careful labeling strategies and normalization algorithms. Dye-swap replicates, in which fluorophores are reversed between experimental and control samples, help correct for dye bias. Spike-in controls can normalize for hybridization efficiency. Signal detection algorithms often employ sliding window analyses, Z-score transformations, or peak calling frameworks adapted from expression arrays. The complexity of this data landscape birthed a new generation of bioinformatic tools—early versions of which would later influence the analytics of ChIP-seq. In many ways, ChIP-chip taught the field how to think statistically about chromatin interactions, laying down the computational infrastructure for the genomics revolution.

What ChIP-chip may lack in resolution, it sometimes compensates for in specificity and accessibility. For studies targeting a defined set of promoters or enhancers—especially in organisms with compact genomes—custom arrays offer a cost-effective and analytically tractable solution. For instance, in yeast, where regulatory elements are tightly packed and well-characterized, ChIP-chip arrays can provide nearly comprehensive coverage of functional regions without the overhead of sequencing. Similarly, in studies involving environmental exposure, developmental transitions, or pathway perturbations, ChIP-chip can deliver focused insights with relatively low sample requirements and data complexity.

These attributes make it attractive to resource-limited laboratories, educational settings, and even clinical pilot studies where sequencing may not be feasible or necessary.

The historical footprint of ChIP-chip is deep and significant. Landmark studies using the technique revealed the genome-wide binding patterns of key developmental transcription factors such as Oct4, Sox2, and Nanog in embryonic stem cells. ChIP-chip was instrumental in mapping Polycomb and Trithorax complexes across silent and active chromatin domains, shedding light on the bivalent chromatin model in pluripotency. In cancer research, ChIP-chip profiles of epigenetic marks unveiled aberrant silencing of tumor suppressor genes and global chromatin rearrangements, catalyzing new avenues of epigenetic therapy. These contributions—cemented in high-impact journals and replicated across laboratories—offered a conceptual leap, proving that genome-wide epigenomics was not only possible, but profoundly illuminating.

As sequencing technologies matured and costs dropped, ChIP-seq began to eclipse ChIP-chip in popularity and performance. The flexibility of sequencing to cover any mappable region, its higher resolution, and the absence of hybridization artifacts made it the new standard. Yet, even as ChIP-chip faded from the forefront, it never vanished entirely. In some applications, such as initial antibody screening, validation of locus-specific effects, or comparative studies across archived arrays, ChIP-chip remains an active tool. Moreover, its historical data sets continue to serve as valuable benchmarks against which new findings are validated. These legacy arrays form the backbone of many compendia and atlases, and their reanalysis with modern algorithms often yields fresh insights.

From an educational standpoint, ChIP-chip is uniquely instructive. It introduces students to key principles of hybridization kinetics, probe design, fluorescence detection, and array normalization—all foundational concepts that echo throughout molecular biology. The hands-on nature of the workflow, from DNA labeling to array scanning, offers a tactile understanding of data generation, fostering an appreciation for the nuances behind every data point. For researchers stepping into the field of chromatin biology, beginning with ChIP-chip can provide a solid grounding before scaling up to the higher complexity of sequencing-based approaches.

As the field continues to evolve, ChIP-chip also serves a symbolic function—a reminder that progress in science is not a leap but a continuum. Every new technology builds on the insights, mistakes, and triumphs of its predecessors. ChIP-chip taught us how to scale ChIP beyond PCR. It taught us how to think genome-wide. It brought bioinformatics into the ChIP conversation and created space for the integrative models we use today. Without it, the path to ChIP-seq might have been slower, more fragmented, and less fruitful.

Today, as we stand in the age of single-cell ChIP-seq, CUT&Tag, and multi-omic integration, it's tempting to look back on ChIP-chip as an obsolete curiosity. But to do so is to miss its enduring legacy. Like the early microscopes that first revealed cells, or the stained gels that separated our first PCR products, ChIP-chip is part of our scientific DNA. It represents a turning point where hypothesis met scale, and where chromatin ceased to be an abstract concept and became a mappable terrain.

For any scientist interested in the evolution of genomic technologies—or in mining targeted, cost-effective data in specific contexts—ChIP-chip remains more than a historical artifact. It is a living, breathing methodology with lessons to teach and data yet to offer. As long as the genome continues to whisper its secrets, and as long as there are minds eager to listen, ChIP-chip will remain part of the conversation—a legacy etched in fluorescence, captured on glass, and mapped in memory.

15. GENOME-WIDE BINDING SITE ANALYSIS

The revolution in understanding the genome did not come solely from sequencing DNA. It emerged from decoding the choreography—when, where, and how proteins interact with the genome to dictate its function. Among all tools devised to unravel this dynamic dialogue, ChIP-seq, or Chromatin Immunoprecipitation followed by sequencing, has emerged as the gold standard. It is not merely a method—it is the lens through which regulatory landscapes are revealed and interpreted. Every transcription factor binding, every histone mark, every occupancy pattern inscribed in our chromatin, now has a voice, and that voice is sequenced base by base with ChIP-seq.

The journey begins in the wet lab, where the abstraction of a genome becomes tangible through fragments of crosslinked DNA-protein complexes. The chromatin is carefully sheared—either sonicated into random segments or precisely cut using nucleases—so that the DNA associated with a protein of interest remains intact within a manageable size. Immunoprecipitation isolates these complexes using a highly specific antibody. But it is what follows that transforms ChIP into ChIP-seq: the captured DNA, bound to its guardian proteins, becomes the template for library preparation—a meticulous process of end repair, adapter ligation, and amplification. Each molecule is tagged, indexed, and readied for sequencing. And in this, every fragment is more than a molecule—it is a coordinate, a signal, a genomic fingerprint waiting to be mapped.

Sequencing technologies have evolved dramatically, and with them, the scope and sensitivity of ChIP-seq. The early reliance on short-read sequencers, particularly Illumina platforms, offered high accuracy and depth. These machines became workhorses in research labs, with reads as short as 50 base pairs revealing entire regulatory landscapes when aligned with robust algorithms. As sequencing evolved, long-read technologies such as Oxford Nanopore and PacBio began entering the scene. While still facing challenges with throughput and accuracy in ChIP-seq applications, they offered glimpses into previously inaccessible regions—repeats, structural variants, and long-range interactions. The choice between short and long-read depends on the biological question. For sharp transcription factor peaks, Illumina suffices. For broader domains and complex chromatin states, hybrid approaches may soon redefine standards.

Multiplexing has added a layer of efficiency and complexity to ChIP-seq workflows. Instead of running one sample per lane, researchers now barcode multiple samples with unique adapters, allowing pooled sequencing. This not only reduces costs but also minimizes technical batch effects. The barcodes, incorporated during adapter ligation, must be carefully designed to avoid cross-talk and ensure distinct identification during demultiplexing. In high-throughput projects like ENCODE or modENCODE, multiplexing has enabled genome-wide chromatin mapping across dozens of cell types in parallel, painting comprehensive atlases of regulatory control.

Yet, behind every elegant dataset is the invisible scaffolding of quality control. Input DNA samples—those not subjected to immunoprecipitation—are essential baselines. They allow for normalization and help discern signal from background. In addition, spike-in controls, such as Drosophila chromatin in human ChIP-seq experiments, have become instrumental. These standards enable inter-sample comparisons and correct for variability in immunoprecipitation efficiency or sequencing depth. Their importance cannot be overstated—without normalization, ChIP-seq risks becoming a relative, rather than absolute, assay.

Once the sequencer has done its job, millions of reads emerge. But raw reads are noise until aligned, filtered, and interpreted. Data processing begins with alignment, typically using tools like BWA or Bowtie, that map each read to a reference genome. Poorly mapped reads, duplicates, and adapter sequences are culled, yielding a clean set of high-confidence alignments. Then comes peak calling—the computational heart of ChIP-seq. Algorithms like MACS2 or SICER scan the genome to find regions where reads cluster significantly above background. A sharp, narrow peak often signifies transcription factor binding; a broad, diffuse signal may indicate histone modifications or chromatin remodelers. Each peak is an event, a footprint, a molecular decision recorded in the chromatin.

Visualizing these peaks transforms data into insight. Genome browsers such as IGV or UCSC Genome Browser allow researchers to overlay tracks of ChIP-seq data with genes, enhancers, and epigenetic marks. The patterns speak volumes: high peaks at promoter regions suggest active transcription; valleys in heterochromatin whisper of repression. Readers become cartographers, mapping valleys and mountains of regulatory topography.

But interpretation goes further than peaks. Peak shape, height, and distribution yield biological meaning. For instance, a bimodal peak centered on a transcription start site might indicate RNA Polymerase II pausing. A series of evenly spaced peaks along gene bodies may reflect nucleosome positioning. Integration with RNA-seq, ATAC-seq, or Hi-C can unveil how these binding events shape transcription, accessibility, and chromatin architecture. ChIP-seq data is rarely standalone. It thrives in a symphony of omics, with each technique playing its note in the grand composition of genome regulation.

Applications of ChIP-seq are as vast as biology itself. In developmental biology, ChIP-seq has revealed the temporal choreography of histone marks as stem cells differentiate. Bivalent domains—regions marked by both H3K4me3 and H3K27me3—unveiled poised genes waiting for activation or silencing. In cancer research, oncogenic transcription factors like MYC or mutated p53 leave unique binding footprints, offering insights into disease mechanisms and potential drug targets. In immunology, enhancer landscapes have been charted across B cells, T cells, and macrophages, unveiling how identity and memory are epigenetically etched. Even in neurobiology, ChIP-seq helps decode how histone acetylation patterns influence learning and memory.

Beyond biology, ChIP-seq has catalyzed new discoveries in medicine. Regulatory mutations, once invisible in exomes, are now interpretable through enhancer disruptions revealed by ChIP-seq. In pharmacogenomics, drug responses are tied to the activation of specific genomic regions. In infectious disease, host-pathogen interactions leave chromatin imprints, revealing immune evasion strategies. The method has become indispensable not only in basic science but in translational pipelines, biomarker discovery, and precision medicine.

But power demands responsibility. Experimental design in ChIP-seq is both art and science. The antibody must be rigorously validated—not just ChIP-grade but batch-tested. Controls, both positive and negative, are non-negotiable. Replicates—biological, not just technical—are essential for reproducibility. Sequencing depth must match the protein of interest; 10 million reads may suffice for histone marks, but transcription factors might demand 50 million or more. Fragment size, GC bias, PCR duplicates—all must be considered in crafting a dataset that is not just publishable, but reliable.

At the end of the pipeline, when peaks are called and tracks are rendered, a final truth emerges. Every peak is a story. A story of a protein that chose to bind, at a specific time, in a specific cell, to control a specific fate. ChIP-seq captures these moments—millions of them—in parallel. And in doing so, it captures the essence of regulation: that the genome is not static, but spoken. Not merely encoded, but enacted. Each read is a whisper of biological memory. Each peak, a declaration of intent.

The genome is not a script—it is a stage. And ChIP-seq reveals its actors, their marks, and their movements. From the silence of a repressed gene to the crescendo of an activated enhancer, this technology translates chromatin dynamics into readable narratives. It is a tool, a window, a revolution in understanding how life is orchestrated at the molecular level.

And so, ChIP-seq becomes more than a method. It is a language—the language of control, memory, and identity—inscribed in chromatin and decoded through science.

16. HIGH-RESOLUTION TECHNIQUES

In the relentless pursuit of greater precision and clarity in chromatin immunoprecipitation (ChIP), two techniques have emerged that elevate the resolution of protein-DNA interaction mapping to near base-pair accuracy—ChIP-exo and ChIP-nexus. These methodologies represent not just incremental improvements but a conceptual leap beyond traditional ChIP and even beyond ChIP-seq. With their capacity to reveal transcription factor footprints and binding dynamics in exquisite detail, they have become essential tools in the modern regulatory genomics toolkit.

ChIP-exo—short for chromatin immunoprecipitation followed by exonuclease digestion—begins similarly to conventional ChIP, relying on crosslinking, sonication, immunoprecipitation, and purification. However, after the DNA is pulled down with the bound protein, a critical refinement step is introduced: lambda exonuclease digestion. This enzyme trims DNA in a 5' to 3' direction but cannot digest past a protein-DNA crosslinked site. The result is a DNA fragment that is shaved right up to the footprint of the bound protein, stopping precisely where the protein obstructs further digestion. By sequencing the ends of these trimmed fragments, researchers can identify the exact base-pair where the protein interacts with DNA. This means no more broad peaks and guesswork; ChIP-exo provides a sharply defined boundary around each binding event, allowing for footprint-level resolution.

ChIP-nexus takes this approach a step further. It integrates a clever use of molecular barcoding and specialized ligation steps to reduce noise and increase signal specificity. After end repair, a unique barcode is ligated to the DNA ends, allowing precise tracking of individual molecules and eliminating PCR duplicates from data analysis. This improves quantification and fidelity, making ChIP-nexus an even more powerful variant for studying low-abundance proteins or binding events in complex chromatin environments. Moreover, the protocol adds an adapter ligation and circularization step that helps prevent artifacts and enables even more accurate mapping of the exact nucleotide positions involved in protein-DNA interactions.

The power of these high-resolution techniques lies in what they reveal. Traditional ChIP-seq tells us where a protein generally binds across the genome—often across a 200 to 500 base pair region. In contrast, ChIP-exo and ChIP-nexus tell us where, precisely, the protein touches the DNA—at single-nucleotide resolution. This allows researchers to see motif alignment and spacing, distinguish between direct and indirect binding, and resolve binding at promoters, enhancers, insulators, and other regulatory elements in unprecedented detail. It also uncovers subtle patterns, such as cooperative binding between multiple proteins, transcriptional pausing signals, and the nuances of complex chromatin architecture.

To achieve such accuracy, however, one must pay the price of stringency. ChIP-exo and ChIP-nexus demand a higher level of input quality and technical precision than their traditional counterparts. Crosslinking must be carefully optimized to preserve epitope integrity while allowing efficient trimming. Immunoprecipitation must be exceptionally clean, since background DNA will also be trimmed and sequenced, potentially amplifying artifacts. DNA recovery and purification must be gentle and efficient, minimizing loss while preserving fragile ends necessary for adapter ligation. Library preparation protocols must be followed with discipline, as any misstep can introduce noise or distort the footprinting signal.

Despite their complexity, these methods are increasingly being applied in key areas of biological research. In developmental biology, they are used to map temporal shifts in promoter occupancy and enhancer activation with stunning precision. In cancer genomics, ChIP-exo reveals how oncogenic transcription factors hijack regulatory elements, often with subtle shifts in motif use or spacing that ChIP-seq cannot resolve. In the study of epigenetics, ChIP-nexus enables fine-scale mapping of chromatin remodeler activity, nucleosome displacement, and cofactor recruitment patterns. Researchers can observe with clarity how different transcription factors co-bind, how insulator elements function at the base-pair level, and how the genome's regulatory logic is encoded in its physical structure.

Moreover, these techniques have contributed to several groundbreaking insights. Studies using ChIP-exo have clarified the architecture of promoter regions in yeast and mammals, showing precise transcription factor positioning relative to TSS (transcription start sites). Others have shown how signal-dependent transcription factors like NF-κB change their

DNA binding landscape in response to cytokines, with shifts in motif preference and cofactor presence visible only with high-resolution techniques. In enhancers, ChIP-nexus has uncovered closely spaced binding sites that would appear as a single peak in ChIP-seq, yet represent distinct protein complexes orchestrating gene activation. These insights have added new dimensions to the way we understand the cis-regulatory code.

When comparing ChIP-exo and ChIP-nexus to ChIP-seq, the difference is akin to upgrading from a coarse-resolution satellite image to a ground-level close-up. ChIP-seq may show a hill, but ChIP-exo reveals the individual stones. Yet both have their place. ChIP-seq remains ideal for broad surveys and large-scale comparisons across conditions, species, or developmental stages. ChIP-exo and ChIP-nexus, by contrast, shine when the question demands precision—when one needs to understand not just if a protein binds, but where it binds, how it binds, and what other factors are involved. Together, they offer complementary views of the regulatory genome.

Protocols for these techniques are demanding but increasingly accessible. Commercial kits now support streamlined workflows for ChIP-exo and ChIP-nexus, including optimized enzymes, barcoding reagents, and validated antibodies. Nonetheless, custom optimization is often necessary. Input chromatin should be fragmented to an optimal size range, usually 100 to 300 bp, to ensure efficient digestion and sequencing. Exonuclease activity must be titrated to avoid over-digestion, and adapter ligation requires clean DNA ends and careful thermal cycling. Pilot experiments and replicates are essential to establish reproducibility and identify batch effects.

Data analysis also demands specialized pipelines. For ChIP-exo, software tools identify the 5' end of reads as the true protein-DNA interaction point, rather than centering on peaks. Duplicate removal, barcode demultiplexing, and motif discovery require refined handling to make the most of the precise information these techniques offer. ChIP-nexus analysis introduces additional complexity, including tracking unique molecular identifiers (UMIs), handling ligation biases, and distinguishing between technical and biological variation. Fortunately, bioinformatics toolsets such as MACE, PIP-seq, and NexTools have been developed specifically to handle this complexity, and are often paired with visualization platforms like IGV or UCSC Genome Browser for footprint-level inspection.

As these high-resolution methods continue to mature, their integration with other genomic assays promises new frontiers. Combining ChIP-nexus with ATAC-seq, for example, allows simultaneous insight into factor binding and chromatin accessibility at single-nucleotide resolution. Pairing with single-cell technologies may one day make it possible to map individual transcription factor footprints in rare cell populations. The integration of spatial genomics and high-resolution ChIP could map not only where factors bind on DNA, but where they bind in physical space within the nucleus. The potential for discovery is vast.

In the context of regulatory genomics, the introduction of ChIP-exo and ChIP-nexus marks a shift toward precision biology. No longer satisfied with coarse maps, researchers now demand regulatory atlases that can pinpoint the molecular levers of gene expression with surgical accuracy. These techniques deliver that clarity, providing a window into the choreography of gene regulation that is as close to the molecular truth as current technology allows.

For those venturing into this frontier, the journey begins with a question—not simply where a protein binds, but how exactly it shapes the genetic narrative written in chromatin. ChIP-exo and ChIP-nexus offer the tools to answer that question with the precision it deserves.

17. EMERGING ALTERNATIVES TO ChIP

In the ever-evolving landscape of molecular biology, new methodologies arise not just to complement existing ones but to revolutionize the very way we interrogate the genome. Chromatin immunoprecipitation (ChIP), long the cornerstone for mapping protein-DNA interactions, has faced growing scrutiny for its limitations—chief among them, high background noise, the requirement for large cell numbers, and reliance on crosslinking and mechanical fragmentation. In this milieu of constraint and compromise, two game-changing innovations—CUT&RUN and CUT&Tag—have emerged to upend the status quo. Elegant in principle and efficient in practice, these methods circumvent the pitfalls of ChIP by adopting a fundamentally different approach to interrogating chromatin. Their rise marks a paradigm shift, promising cleaner data, faster workflows, and broader applicability, even in contexts where ChIP falls short.

At their core, CUT&RUN (Cleavage Under Targets and Release Using Nuclease) and CUT&Tag (Cleavage Under Targets and Tagmentation) leverage enzyme-tethering strategies to precisely cleave DNA at the site of protein binding. Unlike ChIP, which depends on crosslinking DNA to proteins—often irreversibly and imperfectly—these newer techniques take advantage of antibody-guided targeting without requiring such chemical fixation. In CUT&RUN, the micrococcal nuclease (MNase) is fused to protein A (or protein A/G) and directed to the chromatin through specific antibodies. Once at the binding site, the enzyme is activated to cleave nearby DNA, releasing fragments into solution that correspond to the protein's genomic occupancy. CUT&Tag, meanwhile, employs a modified transposase (Tn5), pre-loaded with sequencing adapters. This enzyme not only cuts but simultaneously tags DNA, streamlining the process by integrating library preparation directly into the cleavage step.

The implications of these differences are profound. CUT&RUN and CUT&Tag minimize background signal because they eliminate the need for mechanical shearing, a process that in ChIP can generate nonspecific fragments that muddy downstream analysis. Furthermore, their enzymatic nature allows them to function with dramatically fewer input cells—sometimes as few as 1,000—making them accessible for rare cell populations, primary tissues, and even single-cell applications. CUT&RUN, with its clean signal and robust performance in profiling histone modifications, is particularly adept at charting the epigenetic landscape with precision. CUT&Tag, more versatile in its compatibility with transcription factors, excels in detecting DNA-protein interactions that were previously masked by ChIP's inherent background and resolution limitations.

One of the elegant innovations shared by both techniques is the use of bead-anchoring to tether nuclei or permeabilized cells. In a typical workflow, intact nuclei are immobilized on concanavalin A-coated magnetic beads. The beads serve not only as a scaffold for reactions but also simplify washing and incubation steps, minimizing sample loss. Antibodies specific to the protein of interest are introduced, followed by fusion enzymes—either pA-MNase in CUT&RUN or pA-Tn5 in CUT&Tag. These enzymes remain inactive until a controlled activation step: calcium is used to initiate MNase cleavage in CUT&RUN, while magnesium activates Tn5 in CUT&Tag, promoting adapter integration. The reactions are brief, reducing nonspecific cleavage, and the resulting fragments are readily purified from supernatant or bead fractions.

The execution of these protocols has been further facilitated by the availability of commercial kits and automation-ready platforms. Several biotechnology companies now offer standardized kits for both CUT&RUN and CUT&Tag, incorporating optimized buffers, fusion enzymes, and barcoded adapters for high-throughput compatibility. These kits not only reduce variability between experiments but also democratize access to cutting-edge chromatin profiling for labs lacking specialized equipment. Moreover, the development of single-tube, streamlined versions of the protocols has enabled integration into multi-well formats, opening the door to large-scale screening efforts in functional genomics, cancer biology, and developmental studies.

Data derived from CUT&RUN and CUT&Tag are characteristically sharp, with distinct peak profiles and minimal background. Peak calling is straightforward, and fewer reads are often required to achieve statistical significance compared to ChIP-seq. This efficiency translates into reduced sequencing costs, making these methods particularly attractive for labs operating under tight budgets. Moreover, the data lend themselves well to integration with other epigenomic and transcriptomic datasets, enabling multi-omic analyses with unprecedented resolution. For instance, when combined with RNA-seq or ATAC-seq, CUT&Tag can help dissect transcriptional regulation in dynamic cellular states or drug responses.

Of course, even the most promising technologies are not without challenges. Both CUT&RUN and CUT&Tag demand high-quality antibodies that function in native, non-crosslinked conditions. This requirement can limit the choice of targets, especially for transcription factors with few validated native-binding antibodies. The reactions must be carefully titrated to avoid over-digestion or under-representation of weakly bound regions. In CUT&Tag, there is also a risk of transposase overactivity, which can result in hyper-tagmentation and introduce bias. Nonetheless, these issues are manageable with optimization and controls, and the field continues to improve protocols with newer fusion constructs, enhanced barcoding strategies, and built-in normalization tools.

Comparison charts between ChIP, CUT&RUN, and CUT&Tag starkly illustrate the strengths of the newer techniques. CUT&RUN consistently demonstrates higher signal-to-noise ratios and faster turnaround times. CUT&Tag, with its integrated tagmentation step, reduces library prep to a matter of hours and scales efficiently for multiplexed studies. Both techniques outperform ChIP in sensitivity, often detecting binding events that would otherwise require ultra-deep sequencing. Additionally, the elimination of crosslinking reduces epitope masking and preserves protein conformation, making these methods more physiologically relevant.

In practice, CUT&RUN is now the go-to technique for profiling histone modifications and chromatin states across developmental stages or disease models. Its minimal background and high resolution make it ideal for mapping repressive marks like H3K27me3 or activation marks like H3K4me3. CUT&Tag, on the other hand, has proven powerful in transcription factor studies, especially when coupled with single-cell indexing. Innovations like scCUT&Tag are already enabling researchers to track cell-type–specific chromatin landscapes in heterogeneous tissues, from embryonic development to tumor microenvironments.

In the hands of a skilled researcher, these techniques offer far more than mere alternatives—they represent a new era of chromatin mapping. The ability to profile protein-DNA interactions with such clarity, speed, and efficiency invites a reconsideration of experimental design altogether. Where once ChIP-seq projects required millions of cells and weeks of laborious prep, CUT&RUN and CUT&Tag now make it feasible to generate publication-quality data in a fraction of the time, with a fraction of the input.

The impact of these tools is only just beginning to be felt. As antibody validation expands and fusion constructs evolve, it is likely that the precision and versatility of CUT-based techniques will further improve. Their compatibility with high-throughput workflows, barcoded indexing, and multi-modal analyses ensures their place at the center of future epigenomic discoveries. For researchers committed to understanding the nuances of gene regulation, chromatin architecture, and cellular identity, mastering these techniques is no longer optional—it is essential.

And so, with CUT&RUN and CUT&Tag, the scientific community finds itself on the cusp of a chromatin revolution. These tools are not merely upgrades to ChIP; they are its evolution. They offer a clearer lens through which to observe the genome's inner workings, unclouded by technical noise and enriched by molecular precision. As these methods continue to mature, they will no doubt unlock deeper, more refined maps of the regulatory genome, guiding us closer to the fundamental truths written in the language of chromatin.

18. SINGLE-CELL ChIP AND LOW-INPUT ADAPTATIONS

Imagine holding a single cell in your hand—fragile, microscopic, and brimming with an entire universe of genetic potential. For decades, our understanding of gene regulation and chromatin dynamics has been painted in broad strokes, drawn from populations of millions of cells. But what happens when we look closer—when we peer into the regulatory machinery of just one cell, or a rare subpopulation hiding in the shadows of biological complexity? That is the frontier single-cell chromatin immunoprecipitation (scChIP) and low-input ChIP techniques have boldly entered. This chapter is an exploration of that terrain, where scarcity of material no longer silences discovery.

The birth of single-cell ChIP marks a dramatic leap in molecular biology, not merely in technical achievement but in philosophical shift. Traditional ChIP protocols demanded abundance—millions of cells, vast quantities of chromatin, copious amounts of antibodies—limiting its utility in cases where only minute biological samples were available. But the biological world does not always offer abundance. A tumor biopsy, a preimplantation embryo, a circulating tumor cell, or a neuronal subtype—all of these are precious, irreplaceable, and available in sparse quantities. In these rare situations, every molecule counts. Researchers needed to rethink everything: how chromatin is isolated, how antibodies are used, how signals are preserved and amplified without distortion. In that crucible of need, single-cell and low-input ChIP methods were forged.

At the core of these approaches is innovation in barcoding, droplet-based microfluidics, and indexing strategies. Each cell must be treated not merely as part of a pool, but as an individual witness to its epigenetic history. To accomplish this, scientists harnessed the power of unique molecular barcodes—short DNA sequences that tag chromatin fragments from single cells during library preparation. Once the barcoded DNA is pooled and sequenced, these barcodes act as name tags, telling us exactly which cell each signal came from. This allowed high-throughput profiling of histone modifications and protein-DNA interactions, not just in bulk populations, but within individual cells—each with its own story.

Microfluidic systems have been particularly transformative. These miniaturized lab-on-a-chip devices can isolate single cells into nanoliter-scale reaction chambers or droplets. Within these microenvironments, chromatin is extracted, immunoprecipitated, and barcoded. By handling such small volumes with incredible precision, microfluidics reduces reagent costs, minimizes sample loss, and maintains the integrity of single-cell assays. Platforms like Drop-ChIP and μChIP (micro-ChIP) exemplify this principle in action. Drop-ChIP, for example, encapsulates single cells in droplets with barcoded beads and performs the entire ChIP-seq pipeline within these emulsions. μChIP, on the other hand, scales down the conventional protocol for extremely low-input samples, down to as few as 100 cells or even less.

But working with low-input material is not simply a matter of scaling down. It presents a labyrinth of technical challenges. Chromatin loss, nonspecific binding, antibody variability, and amplification bias become exponentially more problematic as input diminishes. Researchers have responded with a variety of innovations. Carrier chromatin, often from exogenous species like *Drosophila*, can be added to low-input samples to stabilize reactions and provide a spike-in control. Pre-amplification of DNA, once considered a necessary evil, is now executed with extraordinary fidelity through optimized ligation and polymerase chain reactions. New indexing strategies, like combinatorial barcoding, enable thousands of single cells to be processed simultaneously while retaining their individuality.

A major breakthrough came with indexed ChIP-seq strategies. Instead of processing each cell in isolation, scientists can process pools of cells together while maintaining unique barcodes at different steps—first during cell lysis, then during chromatin fragmentation, and finally during library preparation. This combinatorial indexing drastically reduces labor and cost, enabling single-cell epigenomic atlases that were once unthinkable. Technologies such as sci-ChIP-seq (single-cell combinatorial indexed ChIP-seq) and CUT&Tag-based single-cell modifications have emerged from this principle, proving remarkably efficient at capturing histone landscapes in rare or hard-to-access samples.

And what of real-world applications? The insights gleaned from single-cell ChIP are not just academic; they are revolutionary. In cancer research, tumor heterogeneity has long frustrated diagnosis and treatment. A biopsy may yield a dominant population, but what about the rare subclones that drive metastasis or resist chemotherapy? Single-cell ChIP allows researchers to dissect these variations at the chromatin level, revealing transcriptional drivers of malignancy hidden in the noise of bulk analysis. In developmental biology, it illuminates the precise sequence of chromatin remodeling events as a stem cell commits to a specific fate—data that could fuel regenerative medicine and reprogramming strategies.

Take, for instance, a study of early embryonic development. Researchers applied single-cell ChIP to chart the dynamic regulation of H3K4me3 and H3K27ac across the first few cell divisions after fertilization. Despite the near-invisible scale of their input, they observed finely tuned waves of chromatin accessibility and transcription factor occupancy, each marking a developmental checkpoint. Without single-cell resolution, these transitions would blur into obscurity, lost in the averaging of cell populations.

The medical applications are profound. In clinical biopsies—where every cell matters—low-input ChIP techniques allow epigenetic profiling even when sample amounts are vanishingly small. Similarly, fluorescence-activated cell sorting (FACS) can isolate specific subpopulations from blood, tumors, or tissue, which can then undergo ChIP with optimized, picogram-scale protocols. These refinements are not merely academic—they are the groundwork of precision medicine. Imagine tailoring treatment not just to a patient's genetics, but to the precise chromatin landscape of their most aggressive cancer clone, or their rarest immune cell subtype.

Yet the field is not without its limitations. Signal-to-noise ratios remain a challenge. Single-cell data is inherently sparse, with each cell yielding only a fragmentary view of its chromatin state. Aggregating similar cells can provide clarity, but risks losing the very heterogeneity the method was designed to uncover. Antibody quality is paramount, and poor specificity can derail an entire experiment. Moreover, reproducibility remains a concern—what works in one lab may not easily transfer to another without meticulous optimization. There is also the issue of resolution: though we can now map certain histone marks and protein-DNA interactions in single cells, truly base-pair level resolution, especially for transcription factors, remains elusive.

Despite these hurdles, the integration of single-cell ChIP with other omics technologies offers a rich future. Multi-modal platforms that pair single-cell ChIP with RNA-seq, ATAC-seq, or even proteomics are beginning to emerge. These integrated profiles offer a holistic view of cellular identity—one that includes not just what a cell is expressing, but how its chromatin is wired to permit or restrict that expression. Such data may one day allow researchers to predict cell fate transitions, diagnose diseases earlier, and design interventions that reprogram cells at their epigenetic root.

The implications for human health are staggering. In autoimmune diseases, subtle shifts in chromatin architecture within a rogue T-cell population might be the difference between tolerance and attack. In neurodegenerative disorders, early chromatin changes could prelude neuronal loss by years. In infectious disease, understanding how chromatin reprogramming shapes the immune response could redefine vaccine design. And in cancer, where a single mutated cell may seed relapse, single-cell ChIP might offer the ultimate surveillance tool.

As we draw this chapter to a close, the vision becomes clear: a future in which every cell's epigenetic landscape can be charted, compared, and understood. No longer must we settle for averages. No longer must we guess at the secrets buried in rare or complex samples. Single-cell ChIP and low-input adaptations have lifted the veil. The age of cellular individuality is upon us—and with it, a new era of discovery, diagnosis, and treatment.

In the end, it's not just about pushing the limits of technology. It's about seeing biology as it truly is: dynamic, heterogeneous, and exquisitely precise. And with each single cell we decode, we come one step closer to mastering the language of life.

19. ChIP-seq DATA PREPROCESSING AND QUALITY ASSESSMENT

The journey from raw sequencing reads to meaningful biological insight begins not in the peaks or pathways, but in the quiet, rigorous world of data preprocessing. Here, in the digital dust of ChIP-seq experiments, lies the foundation of everything to follow. Before the researcher can speak of regulatory landscapes, transcription factor binding, or histone modification patterns, they must first contend with the chaotic output of sequencers—millions of short DNA fragments, barcoded, error-prone, and raw. Without meticulous handling, even the most beautifully designed ChIP experiment can crumble under the weight of noise, artifacts, and ambiguity.

It begins with the fastq file, that humble container of reads and quality scores. Each read is a short glimpse into the genomic fragments pulled down during immunoprecipitation, and its associated quality string is a coded whisper about how trustworthy that glimpse is. The first task is to look—not to analyze, not yet—but simply to observe. Tools like **FASTQC** become invaluable here. They shine a light on the structure of the data: where adapter sequences still linger, where per-base quality drops, where duplication or GC bias creeps in. With colorful graphs and concise summaries, FASTQC offers a critical first impression, one that reveals whether the sequencing run has delivered on its promise or needs triage.

From this diagnostic phase, action follows. Low-quality bases, especially those trailing at the 3' ends of reads, must be trimmed. Adapter contamination—a common artifact when insert sizes are shorter than read lengths—must be removed with precision. This is where **Trimmomatic** and similar tools like **Cutadapt** enter the scene, allowing researchers to surgically excise unwanted segments without losing valuable data. The trimming step is not glamorous, but it is essential. The cleaner the input, the smoother the downstream alignment, and the more reliable the final interpretation.

Once the reads are trimmed and polished, the next task is to anchor them back to the genome. This alignment step transforms random-seeming sequences into positional information. Tools like **Bowtie2** or **BWA** are used to map each read to a reference genome—often human (hg19 or hg38), mouse (mm10), or other model organisms—depending on the experimental design. Bowtie2, in particular, is favored in ChIP-seq for its balance between speed and sensitivity. But even here, decisions must be made: how many mismatches to allow? What to do with reads that map to multiple locations? ChIP-seq frequently targets repetitive regions, and alignment strategies must walk the tightrope between inclusiveness and specificity.

Following alignment, the data is transformed into more manageable and informative formats. The **SAM** file—Sequence Alignment/Map—is the textual record of every read's placement, rich but unwieldy. It is swiftly converted into the more compact and binary **BAM** format, which supports sorting, indexing, and rapid querying. From here, further transformation into **BED** format may follow, especially for visualizations or region-based analyses. Understanding these file types is not just a matter of syntax—it's about learning the language in which genomic stories are written. The BAM file is the novel; the BED file is the table of contents. Each tells a different story about the same data.

But data quality cannot be assumed just because alignment has succeeded. A second round of assessment now takes place. What proportion of reads aligned to the genome? What fraction aligned uniquely? How many were duplicated? These metrics—**mapping efficiency**, **read depth**, **duplication rate**—are the early warning indicators of ChIP-seq health. High duplication might indicate over-amplification or PCR bias; low mapping rates might signal contamination, poor antibody pull-down, or sequencing failure. Each metric must be interpreted in context. For instance, certain histone marks produce broad, diffuse signals that may seem low in coverage but are biologically appropriate. Others, like transcription factor binding, generate sharp peaks and benefit from deeper sequencing and better resolution.

Contamination, too, can creep in unexpectedly. Reads mapping to unexpected species—*E. coli*, *yeast*, or human in a mouse sample—can suggest barcode leakage or sample mix-up. Low-complexity sequences, too frequent to be biologically meaningful, may indicate artifacts from library preparation. Tools like **MultiQC**, which aggregate quality reports across multiple samples, offer a panoramic view of data integrity. They help ensure that no sample in a batch falls below the standards required for reproducible, trustworthy science.

At this point, normalization emerges as a central theme. Not all ChIP-seq samples are created equal. Different sequencing depths, immunoprecipitation efficiencies, or experimental conditions can bias results. To compare across samples, normalization must be robust. One widely adopted method involves **spike-in controls**—exogenous chromatin, often from *Drosophila melanogaster*, added in known quantities to each sample. Since the same antibody should pull down similar amounts of spike-in chromatin across experiments, variations in spike-in signal reflect technical noise. By scaling sample data relative to the spike-in, researchers can correct for batch effects and ensure more accurate comparisons.

Even without spike-ins, **input control libraries** are essential. These libraries are prepared from the same chromatin but without immunoprecipitation, representing background DNA fragmentation and sequencing bias. Comparing the ChIP sample against its input helps distinguish true signal from random noise. For instance, a peak found in both ChIP and input may reflect open chromatin or tagmentation bias, not a specific binding event. Conversely, a peak enriched in the ChIP library but absent in input is more likely to be genuine.

Consistency across batches is another cornerstone of sound ChIP-seq design. Batch effects—subtle differences in reagents, operators, machines, or sequencing runs—can masquerade as biological variation. To guard against this, **metadata annotation** becomes critical. Every sample should be labeled with its experimental conditions, batch identifiers, library preparation dates, and any deviations from protocol. This metadata, while often neglected, is the compass that allows researchers to navigate and interpret complex data landscapes. Without it, even the most sophisticated analyses risk going astray.

Preprocessing also demands decisions about which reads to retain. Should duplicates be removed? For some marks with high coverage and low expected complexity, like H3K4me3, deduplication may improve specificity. For others, such as transcription factors with naturally sparse signal, removing duplicates may discard true positives. The decision, like many in ChIP-seq, must be informed by the biological context, not merely the computational default.

Eventually, the clean, aligned, and quality-assessed data stands ready for the next steps—peak calling, motif discovery, and biological interpretation. But this readiness is no accident. It is the product of dozens of decisions, each made with care. Every trimming threshold, every alignment parameter, every filtering step ripples forward into the final result. The trustworthiness of a ChIP-seq dataset is not earned at the peak level—it is built here, in preprocessing.

And so this chapter serves as both guide and warning. ChIP-seq is not a push-button technology. It is a craft. Those who wield it must understand not only the biology but the informatics, the statistics, the sequencing intricacies that lie beneath the surface. The raw reads are like marble blocks—full of potential but rough and unshaped. Only through careful sculpting can they become data that tells a story, reveals a mechanism, or illuminates a pathway.

In the end, quality assessment is not a chore—it is a safeguard. It protects the scientific process from false leads, from misinterpretation, and from wasted effort. It ensures that when a peak emerges on a browser track, it does so with the full weight of confidence behind it. For in the world of ChIP-seq, clarity begins not in the biology, but in the preparation. And the most revealing discoveries often start with the cleanest reads.

20. PEAK CALLING ALGORITHMS AND FALSE DISCOVERY RATES

The heart of ChIP-seq analysis lies in a deceptively simple yet profoundly complex task: identifying where proteins bind to DNA and where histone modifications decorate the chromatin landscape. This process, known as peak calling, transforms raw aligned reads into interpretable genomic regions that signal biological meaning. It is here that the true signal emerges from the ocean of sequencing data, yet lurking beneath that clarity is a delicate balance between sensitivity and specificity, discovery and deception. Understanding the algorithms that call peaks—and the statistical rigor they employ to separate real binding events from noise—is essential for any researcher venturing into the world of epigenomics.

At first glance, peak calling might seem straightforward. A peak is simply a genomic region with a higher density of aligned reads than the surrounding background. But biological data is messy. Sequencing depth varies, antibodies have different affinities, chromatin accessibility fluctuates, and biases inherent to sequencing platforms blur the picture. How, then, can one distinguish genuine protein-DNA interactions from random fluctuations or artifacts? This is where peak calling algorithms rise to the challenge, armed with sophisticated statistical models and heuristics that sift through millions of reads to produce lists of candidate binding sites.

One of the most widely adopted tools in this arena is **MACS2** (Model-based Analysis of ChIP-Seq). It is favored for its versatility and robustness, particularly with transcription factor datasets where binding events manifest as sharp, narrow peaks. MACS2 begins by modeling the shift size—the average distance between forward and reverse strand reads—representing the length of DNA fragments immunoprecipitated during ChIP. This shift is critical; it centers the pileup of reads on the true binding site rather than the ends of fragments. By estimating fragment size from the data, MACS2 improves peak localization accuracy.

MACS2 also uses a dynamic Poisson distribution model to predict the expected background read distribution across the genome. It then tests observed read counts against this background to identify regions significantly enriched beyond random chance. Peaks are scored using p-values and q-values (false discovery rate-adjusted p-values), allowing users to set thresholds balancing sensitivity and specificity. The result is a list of peaks, each annotated with statistics indicating confidence and fold enrichment.

Another algorithm, **SPP** (strand cross-correlation peak-finding), takes a different approach by focusing on the correlation between forward and reverse strand reads. SPP calculates cross-correlation profiles to estimate fragment length and peak locations. It has proven effective, especially in transcription factor ChIP-seq, and was widely adopted in ENCODE projects, setting a standard for quality metrics.

For histone modification datasets, which often produce broad, diffuse enrichment patterns rather than sharp peaks, tools like **SICER** (Spatial Clustering for Identification of ChIP-Enriched Regions) offer advantages. SICER clusters enriched regions based on spatial proximity and read density, identifying extended domains rather than pinpoint peaks. This distinction is critical; the biology of histone marks like H3K27me3 or H3K9me3 involves large repressive chromatin domains that defy simple peak definitions. Choosing an appropriate peak caller aligned with the biological mark under study is essential for meaningful results.

While these algorithms are powerful, they rely heavily on parameters that can alter outcomes. P-value thresholds determine statistical stringency—too lax, and false positives flood the output; too strict, and true binding sites might be missed. Fold enrichment, comparing ChIP signal over input control, provides a complementary filter to ensure biological relevance. Fragment size estimation influences peak shape and resolution; inaccurate modeling can blur peaks or fragment broader domains. Understanding the nuances of these parameters and tuning them based on data quality and experimental goals is a critical skill.

Beyond individual peak calling lies the challenge of reproducibility. Biological and technical replicates are the gold standard to ensure findings are not artifacts of a single experiment. But how does one assess reproducibility in peak lists? The **IDR** (Irreproducible Discovery Rate) framework answers this by quantitatively comparing peak rankings between replicates. IDR measures consistency by identifying peaks present across replicates with similar enrichment and rank. Peaks passing IDR thresholds are deemed reproducible, bolstering confidence in downstream analyses. The ENCODE consortium has championed IDR as a best practice, recognizing that reproducibility is the cornerstone of credible epigenomic research.

Eliminating false positives is a continual battle. Noise can arise from chromatin accessibility, PCR artifacts, sequencing biases, or even cross-reactive antibodies. Input controls and IgG controls help establish baseline noise, but residual artifacts persist. Methods to refine peak calling include adjusting parameters, filtering peaks overlapping blacklisted genomic regions known for artifacts, and integrating biological knowledge such as motif presence for transcription factors. Visualization tools—genome browsers like IGV or UCSC—become invaluable here, allowing researchers to inspect peak calls in genomic context, cross-reference with known regulatory elements, and spot inconsistencies.

Poor peak calling not only wastes resources but can distort biological interpretations. False positives may lead to spurious pathway enrichment, mislead hypotheses about gene regulation, or mask critical binding events. Conversely, false negatives can obscure key regulatory mechanisms, stalling discovery. The stakes are high, and precision is paramount.

Modern peak calling is increasingly entwined with machine learning and probabilistic modeling, promising to refine sensitivity and specificity beyond traditional statistical approaches. Yet, these advances complement rather than replace the fundamental understanding of peak calling principles.

At the end of the day, peak calling is the lens through which raw sequencing data is focused into biological insight. Mastery of peak calling algorithms, statistical thresholds, reproducibility metrics, and validation techniques equips researchers with the power to discern true regulatory signals. It transforms mountains of sequencing reads into maps of protein-DNA interaction landscapes—maps that guide the exploration of gene regulation, epigenetic modification, and cellular identity.

In this way, peak calling is more than a computational step; it is the crucible of epigenomic discovery. The peaks it reveals are not just data points—they are the signposts to understanding life's most intricate regulatory networks.

Every peak detected through these algorithms represents a potential regulatory hotspot, a site where transcription factors, co-activators, repressors, or histone modifiers engage the genome to orchestrate gene expression. Yet the complexity of biological systems means that not all peaks are created equal. Some may be transient, cell-type specific, or context-dependent. Others may result from technical artifacts or sequence biases. Distinguishing between these possibilities demands rigorous validation, both computational and experimental.

The biological nature of the target also influences peak calling strategies and interpretation. Transcription factors typically bind discrete, often short DNA motifs, leading to narrow and sharp peaks. Histone modifications, conversely, often span kilobases and form broad domains, necessitating algorithms and parameters tailored to detect diffuse enrichment. The diversity of peak shapes and sizes means that a one-size-fits-all approach is insufficient; a researcher must align peak calling methodology to the underlying biology.

Moreover, the resolution of peak calling is influenced by sequencing depth and fragment size. Higher depth provides greater confidence and finer resolution, but also increases cost and data complexity. Fragment size, estimated during peak calling, affects peak width and localization. Accurate fragment size estimation centers peaks more precisely on binding sites. If underestimated, peaks may appear broad and indistinct; if overestimated, peaks might be fragmented or falsely split.

The interplay between peak shape and algorithm parameters is delicate. For example, MACS2 uses a shifting model that depends on fragment size, optimizing peak detection for transcription factors. For broad marks, SICER employs a clustering approach that defines enriched regions rather than peaks per se. Emerging tools also allow hybrid approaches, capturing complex patterns of chromatin modification.

A vital component in peak calling is the use of control samples, typically input DNA or IgG controls, which account for background signal from sonication bias, open chromatin, or sequencing preferences. Comparing ChIP samples against matched controls enables algorithms to subtract background noise and reduce false positives. This comparison improves the specificity of peak calls and is essential for reliable biological interpretation.

False discovery rates (FDR) are statistical measures estimating the proportion of false positives among all identified peaks. Setting appropriate FDR thresholds is a balancing act; lowering FDR reduces false positives but risks discarding true binding sites, while raising it risks including noise. Most pipelines recommend a 1% FDR cutoff as a starting point, but this may be adjusted depending on experimental design and downstream application.

Beyond single-sample analyses, peak calling gains robustness through replication and statistical reproducibility assessment. Biological replicates capture natural variability, and technical replicates control for experimental noise. The IDR framework quantifies concordance between replicates, providing an objective metric of peak reproducibility. This process often leads to a consensus peak set, representing high-confidence binding sites.

Visualization of peaks in genome browsers serves as both quality control and a hypothesis-generating tool. Overlaying peak calls with gene annotations, known regulatory elements, and other epigenomic datasets provides context and validation. Discrepancies, such as peaks in gene deserts or blacklisted regions, warrant scrutiny and may highlight artifacts.

Errors in peak calling can cascade through analyses, misleading motif discovery, differential binding studies, and integrative multi-omics approaches. Meticulous preprocessing, parameter tuning, replicate analysis, and visualization are the guardrails that prevent such pitfalls.

Looking forward, the integration of peak calling with machine learning and deep learning models promises to enhance sensitivity and specificity by learning complex patterns beyond classical statistics. Models trained on large datasets can predict peak likelihood, annotate functional relevance, and prioritize sites for experimental validation.

In conclusion, peak calling is a fundamental but nuanced step in ChIP-seq data analysis. It transforms raw sequence alignments into meaningful regions of enrichment that decode the regulatory landscape of the genome. Mastering peak calling algorithms, understanding their assumptions, and applying rigorous quality control measures ensure that epigenomic discoveries stand on solid ground. As our tools and datasets grow in complexity, this foundational skill remains central to unlocking the secrets written in chromatin.

21. MOTIF DISCOVERY AND BINDING SITE ANNOTATION

Within the vast landscape of genomic sequences illuminated by ChIP experiments lies a language both ancient and profound—a language written not in words but in recurring patterns of nucleotides known as motifs. These motifs are the cryptic codes that transcription factors and regulatory proteins read and interpret to choreograph the symphony of gene expression. Understanding these motifs, unearthing them from raw binding data, and annotating their genomic contexts are the keys to decoding the regulatory circuits that govern cellular identity, development, and disease.

ChIP data, by its very nature, provides the coordinates of protein-DNA interactions across the genome. Yet, without dissecting the sequences within these regions, the true drivers of binding specificity remain hidden. Motif discovery is the art and science of extracting these underlying patterns—short, recurring nucleotide sequences that represent the binding preferences of transcription factors or their cofactors. This process bridges the gap between observed binding events and the molecular grammar that directs them.

At the forefront of motif discovery are powerful computational tools such as **MEME**, **HOMER**, and **DREME**. Each offers unique algorithms and approaches for revealing motifs from sets of sequences derived from ChIP peaks. **MEME** (Multiple EM for Motif Elicitation) uses expectation-maximization algorithms to identify statistically significant, overrepresented motifs without prior knowledge—this is de novo motif discovery in its purest form. **HOMER** (Hypergeometric Optimization of Motif EnRichment) combines motif discovery with annotation, providing a streamlined workflow tailored for ChIP-seq data, enabling researchers to detect both known and novel motifs. **DREME** (Discriminative Regular Expression Motif Elicitation), designed for speed and specificity, excels at identifying short motifs and is particularly useful when the motif landscape is simple or when rapid results are desired.

The process of de novo motif discovery begins by extracting sequences underlying ChIP peaks—usually a window of several hundred base pairs centered on the summit of enrichment. The computational tools then scan these sequences, comparing nucleotide frequencies at each position to those expected by chance, looking for patterns that recur more often than random distribution. The output motifs are represented as position weight matrices (PWMs), capturing the probability of each nucleotide occurring at each position within the motif. This probabilistic representation reflects the inherent variability in biological binding sites, accommodating degeneracy where a transcription factor may tolerate substitutions without losing affinity.

Contrastingly, motif matching leverages databases of known motifs, such as JASPAR or TRANSFAC, to scan sequences for the presence of established binding patterns. This approach complements de novo discovery by confirming whether observed motifs correspond to characterized transcription factors, or by identifying potential co-factors that may cooperate or compete at the binding site. Together, these methods enrich the biological narrative emerging from ChIP data.

Yet, motif discovery extends beyond isolated sequences. Binding sites often reside within complex regulatory regions where multiple motifs cluster, interacting to form combinatorial codes. Clustering motifs within peaks or across regulatory regions reveals the architecture of transcriptional networks. It uncovers hierarchies where master regulators recruit or inhibit cofactors, sculpting cell-type-specific expression programs. These regulatory modules form the backbone of gene expression control, and their identification provides insight into cellular identity and responses.

The biological significance of discovered motifs is amplified when integrated with evolutionary sequence conservation. Binding sites conserved across species suggest functional importance, preserved by natural selection. Tools that overlay motif locations with conservation scores help prioritize candidate motifs for experimental validation. Conversely, species-specific motifs may highlight regulatory innovations underpinning lineage-specific traits or disease susceptibilities.

Binding site annotation is the natural progression from motif discovery, linking motifs and peaks to genes and functional genomic elements. Annotation tools map peaks to the nearest transcription start sites, enhancers, promoters, or insulators, providing context that transforms abstract sequences into regulatory relationships. This mapping is crucial because not all binding events have equivalent regulatory impact; binding at a promoter may directly influence gene transcription, while distal enhancer binding modulates expression in a context-dependent manner. Accurate annotation enables downstream analyses, including gene ontology enrichment, pathway mapping, and integration with transcriptomic data.

Visualization techniques empower researchers to interpret and communicate motif data. **Motif logos** graphically represent PWMs, depicting the relative frequency of nucleotides at each position and highlighting conserved cores essential for binding. These logos serve as intuitive fingerprints of transcription factors, allowing comparisons between discovered motifs and known binding preferences. Heatmaps, plotting motif enrichment across different peak sets or conditions, reveal patterns of co-occurrence and differential regulation, guiding hypotheses about dynamic regulatory changes.

Critical to robust motif analysis is the use of appropriate background sequences and control datasets. Since genomic sequences exhibit inherent nucleotide composition biases, failing to correct for these can inflate false positives. Background sets often comprise random genomic regions matched for GC content and repeat elements, ensuring that observed motif enrichments reflect true biological signals rather than compositional artifacts. Control datasets, including input DNA or unrelated ChIP samples, further refine motif discovery by subtracting nonspecific enrichments.

Numerous case studies illustrate the power of motif discovery in advancing biological understanding. For example, motif analysis in immune cell ChIP datasets revealed novel cofactor motifs that modulate inflammatory gene expression, uncovering previously unrecognized regulatory circuits. In cancer epigenomics, motif discovery identified aberrant transcription factor binding patterns driving oncogene activation and therapy resistance, providing avenues for targeted intervention. Developmental studies leveraged motif clustering to delineate transcriptional hierarchies guiding lineage commitment and organogenesis.

Beyond static analysis, integrating motif data with chromatin accessibility, histone modification, and gene expression profiles offers a multidimensional view of regulation. It allows researchers to distinguish between potential and actual binding events, infer functional outcomes, and model gene regulatory networks with unprecedented depth. Such integrative approaches propel the field towards a systems-level understanding of gene regulation.

In sum, motif discovery and binding site annotation transform raw ChIP data into a coherent regulatory lexicon. They decode the language through which cells interpret genetic information, respond to environmental cues, and execute developmental programs. Mastery of these techniques equips researchers to uncover the fundamental principles of transcriptional control, elucidate disease mechanisms, and design interventions that modulate gene expression with precision. This chapter serves as a bridge, guiding the reader from data to discovery, from sequence patterns to biological insight, illuminating the pathways that connect DNA to destiny.

As one delves deeper into motif discovery, the intricate balance between sensitivity and specificity emerges as a recurring theme. The algorithms must be sensitive enough to detect true binding motifs, which may be present at low frequency or embedded within noisy data, yet specific enough to avoid spurious motifs arising from random sequence patterns or experimental artifacts. This balance is finely tuned by parameters such as motif length, number of motifs to discover, and statistical thresholds that govern motif significance. Iterative analyses, coupled with biological knowledge, often guide the optimization of these parameters, ensuring that the motifs reported are both biologically plausible and statistically robust.

Moreover, the dynamic nature of transcription factor binding complicates motif discovery. Binding affinity is not solely determined by DNA sequence; it is influenced by chromatin accessibility, DNA methylation, cofactor presence, and three-dimensional genome organization. Consequently, motifs uncovered in ChIP datasets represent not only the preferred binding sequences but also the integrated effect of cellular context. Understanding this complexity requires coupling motif discovery with orthogonal data types—such as ATAC-seq for accessibility or methylome maps for epigenetic modifications—to discern the regulatory logic that governs binding site selection.

Advancements in single-cell technologies have begun to push motif discovery into unprecedented territory. Single-cell ChIP and ATAC-seq datasets reveal heterogeneity in motif usage across cell populations, enabling the identification of cell-type-specific regulatory elements and transcription factor networks. These approaches hold promise for unraveling developmental trajectories, disease progression, and cellular responses to stimuli at an unparalleled resolution.

The interpretation of motifs also benefits from machine learning and deep learning frameworks. Neural networks trained on large compendia of binding data can predict binding affinity and identify subtle sequence features beyond canonical motifs. These models can integrate DNA sequence with epigenetic features, enhancing motif discovery accuracy and enabling predictions in untested cell types or conditions. While computationally intensive, these approaches represent the frontier of motif analysis and hold transformative potential.

Ultimately, motif discovery and binding site annotation are not ends in themselves but tools to uncover the grammar of gene regulation. They reveal the 'words' and 'sentences' that direct gene expression programs, mediate cellular identity, and dictate phenotypic outcomes. Through their application, researchers gain access to the regulatory code written in DNA, opening avenues for novel therapeutics, precision medicine, and synthetic biology.

By bridging raw binding data with regulatory insights, motif analysis empowers the scientific community to translate genome-wide occupancy maps into functional understanding. It is a journey from sequence to function, from data points to networks, illuminating the fundamental mechanisms that underlie life itself.

22. INTEGRATING ChIP DATA WITH TRANSCRIPTOMICS AND EPIGENOMICS

In the intricate dance of gene regulation, no single player acts in isolation. Chromatin immunoprecipitation (ChIP) data, rich with information about protein-DNA interactions and histone modifications, offers a vital perspective, yet it captures only a slice of a far more complex biological narrative. To truly comprehend how genes are turned on or off, how cellular identity is forged, or how aberrations in regulation lead to disease, ChIP data must be woven together with other layers of genomic and epigenomic information. This integration transforms isolated observations into holistic portraits, enabling researchers to decipher the symphony of regulation that orchestrates life at a molecular level.

Central to this integrative approach is the correlation between ChIP peaks and gene expression profiles derived from RNA sequencing (RNA-seq). While ChIP identifies where transcription factors or histone marks bind along the genome, RNA-seq quantifies the functional outcome—transcripts produced or repressed in a given condition. Linking these two data types allows for the identification of direct transcriptional targets, illuminating which binding events translate into changes in gene activity. The power of this correlation lies in its ability to move from static occupancy maps toward dynamic regulatory networks, revealing cause and effect within the genome.

However, bridging ChIP and RNA-seq data is not trivial. Peaks often reside far from annotated genes, within enhancers or distal regulatory elements, complicating direct association. Sophisticated computational frameworks map peaks to genes based on genomic proximity, chromatin loops, or enhancer-promoter interactions identified through techniques like Hi-C. These mappings enrich interpretation, permitting the construction of gene regulatory networks that reflect true biological connectivity.

Beyond expression, the physical accessibility of chromatin at ChIP binding sites adds a crucial dimension. Techniques such as Assay for Transposase-Accessible Chromatin using sequencing (ATAC-seq) and DNase I hypersensitive site sequencing (DNase-seq) reveal open chromatin regions permissive for factor binding. Integrating these datasets with ChIP confirms that observed protein-DNA interactions occur within accessible genomic contexts, strengthening confidence in the biological relevance of identified sites. Conversely, binding events detected in inaccessible chromatin may indicate poised or transient interactions, inviting further scrutiny.

The regulatory landscape is further complicated by DNA methylation, a covalent modification that often acts as a repressive mark. DNA methylation profiling, when integrated with ChIP data, uncovers the interplay between transcription factor binding, histone modifications, and methylation states. For example, hypomethylated regions enriched with activating histone marks and bound by transcription factors define active regulatory elements, while hypermethylated domains may correlate with repression. This triad of information paints a nuanced picture of epigenetic control.

To synthesize these multifaceted datasets, specialized computational tools are indispensable. Software like **DESeq2** enables robust differential gene expression analysis from RNA-seq, providing quantitative measures of transcript changes. **DiffBind** facilitates differential binding analysis between ChIP samples, highlighting condition-specific binding events. **ChromHMM** integrates multiple histone marks to segment the genome into functional chromatin states, illuminating promoter, enhancer, insulator, and repressive domains. By combining these tools, researchers gain the capacity to build comprehensive models of gene regulation that span sequence, occupancy, chromatin environment, and expression output.

An essential aspect of successful data integration is experimental design, emphasizing matched time points and consistent sample conditions across omics layers. Without temporal and biological concordance, correlating datasets risks confounding factors and spurious conclusions. When done carefully, integrated analyses uncover dynamic regulatory processes—for example, how transcription factor binding at an enhancer precedes chromatin opening and subsequent target gene activation during differentiation.

Visualization is key to interpreting the complex data matrices generated by multi-omics integration. Genome browsers enriched with layered tracks for ChIP, ATAC-seq, DNA methylation, and RNA expression allow intuitive exploration of regulatory landscapes. Heatmaps and cluster analyses reveal patterns of co-regulation and identify gene modules controlled by common regulatory factors. Network diagrams map interactions between transcription factors, cofactors, and their gene targets, illuminating the architecture of gene regulatory networks. These visual tools not only facilitate discovery but also enhance communication of findings across interdisciplinary teams.

The integration of ChIP with transcriptomics and epigenomics has already yielded transformative insights. In developmental biology, such approaches have unraveled the stepwise activation and repression of gene programs that guide cell fate decisions. In cancer research, integrated datasets have delineated aberrant regulatory circuits driving oncogenesis, metastasis, and therapeutic resistance. In immunology, multi-omics reveal how chromatin remodeling coordinates immune cell activation and memory formation. Each example underscores the power of data integration to reveal biological truths inaccessible through isolated analyses.

Yet challenges remain. Data heterogeneity, differences in resolution and sensitivity, and batch effects can complicate integration. Computational frameworks continue to evolve, aiming to harmonize diverse data types and extract biologically meaningful patterns. Machine learning approaches are increasingly applied to multi-omic datasets, enabling predictive modeling of regulatory outcomes and identification of novel regulatory elements.

Ultimately, integrating ChIP data with transcriptomics and epigenomics shifts the analytical paradigm from single-layer observations to systems-level understanding. It transforms raw datasets into narratives about cellular states, regulatory hierarchies, and molecular mechanisms. This holistic perspective empowers researchers to tackle complex biological questions, ranging from understanding fundamental gene regulation to developing targeted therapeutics.

As we continue to expand the depth and breadth of omics technologies, the integration of ChIP with other data modalities will become increasingly essential. The promise lies not just in generating data but in weaving it into coherent stories that illuminate the code of life. This chapter serves as a guide for navigating the exciting frontier of multi-omic integration, equipping readers with the conceptual and practical tools to harness the full power of genomic and epigenomic data toward transformative biological discovery.

The path to truly comprehensive integration is marked by careful attention to detail and a deep understanding of each data type's strengths and limitations. ChIP-seq, while powerful for mapping protein-DNA interactions, reflects a snapshot influenced by antibody specificity, sequencing depth, and chromatin context. RNA-seq captures steady-state or dynamic transcript levels, yet is shaped by mRNA stability, splicing, and post-transcriptional regulation. ATAC-seq and DNase-seq reveal chromatin accessibility but do not confirm factor binding per se. DNA methylation provides a relatively stable layer of epigenetic information but can vary dramatically between cell types and developmental stages. Recognizing these nuances enables researchers to interpret integrated datasets with appropriate caution and insight.

One of the most compelling applications of integrated ChIP and transcriptomic data is in the dissection of transcription factor networks. By overlaying binding sites with changes in gene expression, it becomes possible to distinguish direct targets—genes whose promoters or enhancers are bound and whose expression is correspondingly altered—from indirect targets influenced by downstream signaling. This distinction refines our models of gene regulation, enabling the identification of master regulators and their hierarchical effectors within the cell.

Integrative epigenomics further enriches this view. Chromatin states inferred from histone modifications, DNA methylation patterns, and accessibility data define the regulatory landscape in which transcription factors operate. For example, the combination of activating marks like H3K4me3 at promoters with open chromatin and unmethylated DNA typically signifies active transcription, whereas the presence of repressive marks such as H3K27me3 and dense methylation denotes silenced regions. By mapping transcription factor binding onto these landscapes, researchers uncover how binding correlates with chromatin context, revealing mechanisms of activation, repression, or poised states.

The temporal dimension adds yet another layer of complexity and opportunity. Time-course experiments measuring ChIP binding, chromatin accessibility, and gene expression simultaneously allow the reconstruction of regulatory cascades. They capture how changes in transcription factor occupancy lead to chromatin remodeling, which in turn affects transcriptional outcomes. Such dynamic integration has been invaluable in fields like developmental biology and immunology, where cellular states transition rapidly and precisely.

Advances in single-cell technologies are now pushing multi-omic integration to unprecedented resolution. Single-cell RNA-seq paired with single-cell ATAC-seq or ChIP methods reveal heterogeneity in regulatory programs across individual cells, uncovering rare subpopulations and lineage trajectories obscured in bulk analyses. These approaches herald a new era in understanding how gene regulation shapes cell identity and function at the most granular level.

Practical integration requires robust computational infrastructure and expertise. Pipelines that preprocess, normalize, and harmonize data from different platforms are essential. Statistical methods to identify correlated patterns and causative links must be carefully chosen to avoid spurious associations. Visualization tools that can layer multiple data types while remaining interpretable are critical for hypothesis generation and communication.

In sum, the integration of ChIP data with transcriptomics and epigenomics represents a paradigm shift, moving from isolated data silos to interconnected, multi-dimensional views of gene regulation. This holistic perspective not only deepens our fundamental understanding but also enhances our ability to identify biomarkers, understand disease mechanisms, and develop targeted therapies. As technology and analytical methods evolve, this integrative approach will remain central to the future of genomics and precision medicine, illuminating the complex regulatory codes that govern life.

23. VISUALIZATION TOOLS AND GENOME BROWSERS

In the realm of genomic research, data without visualization is like a story without a storyteller—rich with detail but lacking the clarity to engage, convince, or inspire. The ability to visualize chromatin immunoprecipitation (ChIP) data transforms complex numerical and positional information into intuitive images, enabling researchers to explore, interpret, and communicate the intricacies of protein-DNA interactions and epigenetic landscapes. This transformation is not merely aesthetic; it is fundamental to discovering patterns, validating findings, and weaving biological narratives from raw sequences and peaks.

At the forefront of genomic visualization stand powerful platforms such as the Integrative Genomics Viewer (IGV), the University of California Santa Cruz (UCSC) Genome Browser, and the WashU Epigenome Browser. These tools are more than passive viewers; they are interactive portals into the genome that allow researchers to overlay multiple data types, zoom into specific loci, and annotate features with precision and clarity.

IGV, renowned for its speed and flexibility, offers a desktop application that supports a wide array of data formats including BAM, BED, BigWig, and VCF. Its capacity to load multiple tracks simultaneously allows the visualization of ChIP-seq peaks alongside RNA-seq expression profiles, DNA methylation patterns, or other epigenetic marks, providing a comprehensive view of genomic regulation in a single interface. IGV's intuitive controls for zooming, panning, and track manipulation enable users to explore genomic regions from broad chromosomal views down to single-nucleotide resolution. Customizable color schemes, track grouping, and feature highlighting enhance the interpretability of complex datasets.

The UCSC Genome Browser offers a web-based platform with an expansive repository of reference genomes and annotation tracks. Its strength lies in integrating user-uploaded ChIP data with a wealth of public datasets including gene models, conservation scores, known variants, and regulatory elements. Users can create custom tracks to display ChIP peaks, coverage, or differential binding data, and employ color coding to distinguish between conditions, replicates, or modifications. The browser supports visualization of large datasets through optimized formats such as BigWig and BEDGraph, which efficiently handle continuous signal data and discrete peak intervals, respectively. UCSC's suite of tools includes session saving and sharing features, facilitating collaboration and publication.

Complementing these, the WashU Epigenome Browser provides an elegant and highly interactive web interface optimized for epigenomic datasets. It excels in comparative visualization, allowing side-by-side views of multiple samples, time points, or experimental conditions. The browser integrates sophisticated annotation tracks, chromatin state segmentations, and three-dimensional genome conformation data, enabling a multidimensional perspective on ChIP data. Features like dynamic heatmaps, bar charts, and scatter plots embedded within the browser help users to analyze patterns of enrichment and co-localization across genomic regions.

A critical aspect of effective visualization is the ability to create custom tracks tailored to specific datasets and research questions. Researchers learn to prepare and format data into browser-compatible files such as BED for peak regions, BigWig for continuous signal tracks, and BEDGraph for variable coverage. Color coding is a powerful tool here, allowing differentiation between transcription factors, histone marks, treatment conditions, or replicates. Thoughtful use of color and track organization reduces cognitive load and highlights biologically meaningful contrasts.

Heatmaps and coverage plots extend visualization beyond individual loci to genome-wide comparisons. Heatmaps summarize signal intensity across sets of peaks or genes, enabling pattern recognition and clustering of similar regulatory elements. Coverage plots display average enrichment profiles around features such as transcription start sites or enhancers, illustrating characteristic binding patterns or histone modification landscapes. These visual summaries complement genome browsers, providing context and statistical insight.

Good visualization practices also involve careful figure design for communication. Screenshots from genome browsers become figures in publications and presentations, but their effectiveness depends on clarity, appropriate scaling, and annotation. Including coordinate scales, gene names, peak calls, and clear legends ensures that images convey the intended message without ambiguity. Highlighting regions of interest, using arrows or brackets, and avoiding overcrowding are simple yet impactful techniques.

Beyond two-dimensional views, advances in three-dimensional genome visualization are opening new frontiers. Tools like HiGlass enable the exploration of chromatin interactions and topologically associating domains (TADs) alongside ChIP and other epigenomic data. These 3D browsers contextualize binding events within the spatial architecture of the nucleus, offering insights into regulatory mechanisms influenced by genome folding.

Ultimately, visualization is the bridge that connects computational analyses with biological understanding. It turns abstract data points into stories of regulation, interaction, and control. By mastering visualization tools and techniques, researchers not only enhance their capacity to interrogate data but also their ability to communicate discoveries to the scientific community and beyond.

This chapter invites readers to see the genome not just as sequences and numbers, but as a vivid landscape where proteins and modifications paint dynamic patterns of life. It transforms raw ChIP data into compelling, interpretable, and impactful biological narratives—stories that illuminate the hidden choreography of the genome.

Every dataset tells a different story, and the strength of a researcher lies in their ability to read, interpret, and convey those stories with clarity. While the numbers and enrichment values in ChIP experiments contain a wealth of information, their full potential is unlocked only when they are visualized meaningfully. This is why genome browsers are indispensable—not only for exploration but also for hypothesis generation. By loading aligned ChIP-seq data as tracks atop reference genomes, one can immediately discern patterns of occupancy, discover unanticipated binding regions, or detect anomalies that might otherwise be lost in tables of raw output.

Imagine observing a striking peak of H3K27ac at the promoter region of a gene that is highly upregulated in your RNA-seq data. A glance through IGV confirms this correlation, and adjacent tracks reveal open chromatin signals from ATAC-seq and low DNA methylation—together forming a compelling case for active transcription. Without these visual cues, the association might seem statistically interesting; with them, it becomes biologically persuasive.

Custom tracks empower researchers to create these visual narratives with intent. For example, one might overlay transcription factor peaks from different conditions, each in a unique color, allowing side-by-side comparison of dynamic binding changes in response to a stimulus. Differential peak regions can be emphasized with bold strokes or contrasting hues, while background tracks such as gene annotations and histone marks add functional context. This layering effect is akin to geological strata—each track adds a new depth, a new insight into the regulatory landscape.

The choice of file formats is crucial for efficient rendering and compatibility. BigWig files, in particular, are favored for their compressed yet fast-access capabilities. They enable the smooth display of signal intensities across large genomic regions without requiring the entire dataset to be loaded into memory. BEDGraph files serve a similar purpose but are often less efficient with very large datasets. BED files, on the other hand, are the workhorse for displaying regions of interest, such as peak calls or annotations, and can be enriched with score columns or custom labels.

Visual summaries such as heatmaps distill this complexity into intuitive matrices. A heatmap displaying ChIP-seq signals centered on enhancers, clustered by signal intensity, can reveal subtypes of regulatory elements—some poised, others active, others repressed. When overlaid with gene expression data, such as log2 fold change values from DESeq2, the connections between chromatin state and transcriptional outcome become visible. Heatmaps also help validate reproducibility, showing consistent signal patterns across biological replicates or experimental conditions.

Coverage plots further reinforce these insights. By aggregating signal across multiple genomic features, such as all TSSs or all known enhancers, one can observe average binding patterns that characterize a given factor or histone modification. These plots are not just decorative; they reveal mechanistic trends. A sharp spike of H3K4me3 at the +1 nucleosome position relative to the TSS, followed by a broad domain of H3K36me3 into the gene body, illustrates a textbook example of active transcription that is visually encoded in these plots.

High-quality figure design is not a trivial matter—it reflects the rigor and clarity of the research itself. When submitting a manuscript or preparing a presentation, the ability to generate clean, readable, and informative genome browser snapshots is invaluable. It requires attention to resolution, aspect ratio, labeling, and the judicious use of color. Legends should be descriptive but not cluttered. Track names must be clear, units defined, and scales aligned. Poorly designed figures can undermine the strength of otherwise excellent data, while effective visual storytelling can elevate the narrative to new heights.

The power of visualization becomes even more evident in collaborative environments. Genome browsers that allow session sharing, such as UCSC's custom sessions or WashU's saved states, promote teamwork and interdisciplinary interpretation. A biologist may focus on transcription start sites, a computational scientist on coverage profiles, and a clinician on disease-relevant variants—all within the same visual framework. These shared views foster dialogue and drive discoveries that span disciplines.

As the field progresses, new visualization platforms are emerging to handle the scale and complexity of multi-omic datasets. HiGlass, for instance, brings interactive heatmaps of genome-wide interactions into the browser, supporting synchronized views across different tracks and chromosomes. Its ability to handle matrix-based data from Hi-C, Capture-C, or other 3D assays enriches the interpretive landscape. Researchers can now view ChIP-seq peaks not only in the linear genome but also within the folded three-dimensional context where regulatory loops and TAD boundaries play crucial roles.

Ultimately, what sets expert genomicists apart is not just their ability to analyze data but their ability to tell stories through it. A well-annotated browser image showing the binding of a transcription factor at a key enhancer, co-localized with open chromatin, hypomethylation, and elevated expression of a nearby gene, tells a complete story at a glance. It draws the eye, stimulates curiosity, and invites deeper questions: Why here? Why now? What regulates this binding, and what are its consequences?

ChIP data, when visualized well, becomes more than just science—it becomes art. It captures the fluid interplay of regulation in the genome's language and renders it visible, tangible, and comprehensible. In this light, genome browsers are not merely software tools; they are microscopes of the molecular world, windows into the code of life, and canvases upon which the biological symphony of transcription, repression, and modulation is vividly painted.

For any researcher invested in chromatin biology, mastering these tools is not optional—it is essential. The ability to move seamlessly between raw data and visual interpretation, to generate hypotheses from what is seen, and to communicate those insights with elegance and authority, is what distinguishes impactful science. Visualization closes the loop between discovery and dissemination. It empowers scientists to not just find meaning in the genome—but to show it.

24. INTERPRETING REAL ChIP-seq DATASETS

In the vast landscape of genomic technologies, ChIP-seq has emerged not only as a staple of molecular biology but as a window into the inner logic of the cell—its regulations, its silences, and its ruptures. Yet while the protocols, algorithms, and pipelines offer the technical backbone, it is only through real-world application that the true power of ChIP-seq is realized. Case studies—drawn from cancer biology, developmental genomics, and immunology—illustrate the journey from hypothesis to discovery, from raw signal to insight. These stories are not merely examples; they are models of scientific reasoning, where ChIP-seq becomes a narrative medium, uncovering not just where proteins bind, but why.

One of the most striking uses of ChIP-seq has been in the context of cancer epigenetics. In a landmark study on estrogen receptor alpha (ERα) in breast cancer, researchers sought to understand why certain tumors respond to hormonal therapy while others resist. The hypothesis was elegant in its simplicity: resistance might be reflected in differential ERα binding across the genome. ChIP-seq was performed on patient-derived tumor xenografts—one group responsive to tamoxifen, the other resistant. What emerged was a divergent chromatin landscape. In responsive tumors, ERα bound classically to promoter-proximal estrogen response elements (EREs), orchestrating transcriptional programs tied to growth and differentiation. But in resistant tumors, ERα was redistributed—occupying distal enhancer regions co-bound by the transcription factor FOXA1, suggesting a rewiring of the regulatory circuitry. This spatial reallocation of ERα not only explained transcriptional differences but also pointed toward potential co-targets for therapy. Visualizations in the UCSC Genome Browser displayed dramatic contrasts in ERα occupancy at key loci like GREB1 and TFF1, and heatmaps of differential binding further underscored the remodeling of enhancer landscapes. These patterns were not abstract—they were fingerprints of resistance etched in chromatin.

In developmental biology, another study turned to ChIP-seq to probe the epigenetic transitions during early embryogenesis. Researchers isolated cells at successive stages of mouse embryonic stem cell differentiation—from naïve pluripotency through lineage commitment. They performed ChIP-seq for key histone modifications—H3K4me3 (active promoters), H3K27ac (enhancers), and H3K27me3 (repression)—as well as for lineage-specific transcription factors like Oct4, Sox2, and Nanog. Through comparative analysis, they uncovered that bivalent domains—regions marked simultaneously by H3K4me3 and H3K27me3—were a hallmark of poised developmental genes. As differentiation progressed, these domains resolved: genes activated for a given lineage gained H3K27ac and shed repressive marks, while those repressed acquired stronger H3K27me3. ChIP-seq tracks across loci like Pax6, Brachyury, and Gata6 told the story in chromatin. What was once theoretical—that embryonic cells exist in a poised epigenetic state—was now evidenced in peaks, valleys, and transitions. Integrating RNA-seq data further validated these shifts in expression, creating a layered, time-resolved view of differentiation. Such studies exemplify how ChIP-seq transcends static maps and becomes a dynamic chronicle of cellular fate.

In the realm of immunology, ChIP-seq has illuminated the precise choreography of signaling cascades. Consider the transcription factor NF-κB, a central player in immune response. A team set out to dissect how NF-κB coordinates gene expression following TLR (Toll-like receptor) activation in macrophages. They performed time-course ChIP-seq for the p65 subunit of NF-κB, coupled with RNA-seq and ATAC-seq. Within minutes of stimulation, ChIP-seq revealed rapid binding of p65 at canonical immune loci—IL6, TNFα, and CXCL10—with increasing peak intensity corresponding to activation kinetics. Interestingly, p65 also bound latent enhancers—regions previously inaccessible—which showed a gradual increase in ATAC-seq signal. These latent sites, while transcriptionally silent initially, gained H3K27ac over hours, suggesting a feed-forward regulatory mechanism. Importantly, deletion of these enhancers via CRISPR validated their role in modulating late-phase gene expression. The ChIP-seq visualizations—sharp, emergent peaks at both early and delayed timepoints—were matched with transcriptional outputs, mapping stimulus to response with precision. This case illustrated that NF-κB is not a mere switch, but a sculptor of immune kinetics, carving waves of gene expression through a shifting chromatin landscape.

But not all ChIP-seq studies yield crystal-clear insights. Some underscore the method's limitations and the importance of careful experimental design. In a study attempting to map the binding of a low-abundance transcription factor in a rare neuronal subtype, researchers faced signal dropout and high background. The input control library was poorly matched, resulting in excessive false positives during peak calling. Without spike-in normalization, cross-sample comparisons failed to reflect true biological differences. The study served as a cautionary tale—showing how improper antibody validation, insufficient sequencing depth, and noisy input controls could derail interpretations. Yet even here, lessons abounded. The team re-analyzed the data with MACS2 using a narrower bandwidth and more stringent false discovery thresholds, salvaging core peaks. They validated key binding events using qPCR and redesigned follow-up experiments using CUT&Tag for higher sensitivity. The episode exemplified scientific resilience—and reminded readers that every peak in ChIP-seq is not a promise, but a possibility that demands careful proof.

Further case studies explored tissue-specific regulation. In a cross-tissue atlas of enhancer activity, researchers performed H3K27ac ChIP-seq across twenty human tissues—from brain to liver to lung. Despite the same genome in every cell, the resulting enhancer maps were stunningly diverse. Brain-specific enhancers clustered near neurodevelopmental genes like NEUROD2 and SYT1, while hepatic samples revealed enhancers near albumin and cytochrome P450 genes. By applying ChromHMM to integrate histone marks, they built chromatin state maps that defined active promoters, enhancers, and repressed regions in a tissue-specific manner. These maps, when overlaid with eQTL data, revealed how enhancer polymorphisms shaped gene expression in a population context—offering insights into complex traits and disease. The genome, it seemed, was not just a static code but an interpreter of context, modulated by chromatin.

In clinical studies, ChIP-seq has ventured into biopsy-scale inputs and rare pathologies. One case from pediatric oncology examined H3K27me3 landscapes in DIPG (Diffuse Intrinsic Pontine Glioma), a devastating brain tumor driven by mutations in histone H3.3. Researchers hypothesized that the H3K27M mutation globally reduces H3K27me3, altering the epigenetic repression landscape. ChIP-seq confirmed this loss—widespread erosion of H3K27me3 across repressive domains—but with residual focal peaks at critical loci like CDKN2A. These residual marks hinted at potential druggable vulnerabilities. The team combined this with RNA-seq and pharmacological inhibition studies, identifying candidate genes re-silenced by EZH2 inhibition. ChIP-seq here was not just diagnostic but therapeutic—a compass pointing toward clinical strategies.

Across all these studies, certain themes recurred. Hypothesis formulation guided experimental design: transcription factor vs. histone mark, broad vs. narrow peaks, static vs. dynamic states. Proper controls—input DNA, spike-ins, and biological replicates—were the bedrock of interpretability. Visualization made the abstract tangible, turning statistical enrichment into biological relevance. And integration with orthogonal datasets—RNA-seq, ATAC-seq, methylation, and proteomics—transformed individual peaks into regulatory networks. Whether in the glow of success or the grit of troubleshooting, these cases brought ChIP-seq to life.

Ultimately, this chapter is not a catalog of achievements but a map for those charting their own journey through the genome. By walking through these real datasets—seeing their figures, reading their narratives, understanding their decisions—readers gain not only technical fluency but scientific intuition. They learn to see not just peaks, but patterns. Not just noise, but nuance. They come to understand that ChIP-seq is not merely a technique, but a way of listening—to the language of the genome, to the whispers of regulation, and to the coded histories written in chromatin.

25. APPLICATIONS IN CANCER BIOLOGY AND DISEASE RESEARCH

Chromatin Immunoprecipitation, or ChIP, began as a molecular technique nestled within the academic confines of gene regulation research, yet it has since emerged as one of the most incisive tools in deciphering the epigenetic underpinnings of human disease. Nowhere is this more evident than in cancer biology, where the veil between health and malignancy is often woven not only from genetic mutations, but also from aberrant transcriptional control and epigenetic rewiring. The modern oncologist and molecular pathologist increasingly find themselves examining not just mutations in DNA sequences, but the altered architecture of chromatin—how histone modifications, transcription factor occupancy, and enhancer landscapes reshape the identity and fate of a cell gone rogue.

In the intricate domain of cancer, ChIP-seq has been a game-changer. From identifying drivers of tumorigenesis to mapping silenced tumor suppressors, ChIP provides a window into how the genome is functionally interpreted rather than simply encoded. One of the earliest transformative applications of ChIP-seq was in acute myeloid leukemia (AML), where the oncogenic fusion protein MLL-AF9 was found to hijack transcriptional programs by mislocalizing histone methyltransferase activity. This not only altered gene expression patterns but also defined a distinct epigenetic signature that could be exploited diagnostically and therapeutically. Similarly, breast cancer research has leveraged ChIP to understand the transcriptional circuitry driven by estrogen receptor alpha (ERα). By combining ChIP-seq of ERα with RNA-seq and enhancer mapping, researchers were able to reveal how this nuclear hormone receptor orchestrates a wide network of gene expression changes that contribute to both tumor growth and therapeutic resistance.

The power of ChIP lies in its ability to dissect these transcriptional hierarchies with remarkable resolution. For example, in glioblastoma multiforme (GBM), ChIP-seq profiling of histone marks such as H3K27ac and H3K4me3 uncovered super-enhancers associated with genes that define the glioma stem cell state. These enhancer landscapes were not merely decorative features of chromatin; they were functional drivers, conferring aggressive growth and resistance to differentiation therapies. Disruption of these enhancer hubs through epigenetic inhibitors, such as BET bromodomain blockers, reduced tumorigenicity and demonstrated the actionable potential of ChIP-guided insights.

What makes ChIP particularly indispensable in cancer is its compatibility with mutation data. Mutations in chromatin modifiers such as EZH2, a histone methyltransferase that catalyzes H3K27me3, can be mechanistically contextualized through ChIP. In follicular lymphoma and other B-cell malignancies, gain-of-function mutations in EZH2 result in a hyper-repressive chromatin environment that silences critical differentiation genes. ChIP-seq profiles in these contexts not only confirm altered distribution of H3K27me3 but also predict which genes are epigenetically disabled. These insights have paved the way for targeted therapies such as tazemetostat, an EZH2 inhibitor approved for certain cancers. The story of EZH2 exemplifies the arc from ChIP-based discovery to clinical translation—a blueprint increasingly followed in modern oncology.

ChIP is not restricted to histone marks. The method has been essential for validating transcription factor targets implicated in cancer, such as MYC, TP53, and NF-κB. MYC, a notoriously undruggable oncogene, is a master regulator of transcription, binding to thousands of sites across the genome. ChIP-seq has allowed researchers to map these binding events and determine which genes are directly influenced by MYC under oncogenic versus homeostatic conditions. When ChIP-seq is combined with CRISPR screening, the picture becomes even clearer: one can disrupt putative MYC targets and observe phenotypic consequences, validating regulatory networks identified by ChIP. This marriage of epigenomics with functional genomics exemplifies the depth of insight now achievable in cancer research.

Beyond cancer, ChIP has cast new light on autoimmune and neurodegenerative disorders, where dysregulation often stems from subtle epigenetic shifts rather than gross mutations. In systemic lupus erythematosus (SLE), for instance, ChIP-seq of H3K27ac in monocytes has identified abnormal enhancer activation at immune response genes. These aberrant regions correlate with disease severity and response to therapy, suggesting potential roles as biomarkers. In multiple sclerosis (MS), transcription factors such as STAT3 and IRF4 have been profiled through ChIP-seq, revealing immune cell–specific enhancers that contribute to neuroinflammation. This form of precision immunology is now guiding the design of more targeted interventions, particularly those that modulate epigenetic readers and writers.

In neurodegenerative disease, ChIP continues to elucidate the mechanisms by which epigenetic drift contributes to pathology. In Alzheimer's disease, ChIP-seq profiling of brain tissue has revealed loss of H3K27ac at synaptic plasticity genes, coupled with gain of repressive marks like H3K9me3. These epigenetic scars suggest a slow but steady erosion of neuronal identity, providing molecular correlates to clinical cognitive decline. The findings support a growing hypothesis that Alzheimer's may, in part, be an epigenetically mediated disease, opening avenues for intervention before irreversible damage accrues.

One of the more exciting frontiers is the use of ChIP to track the evolution of resistance in cancer therapy. Drug resistance remains a formidable barrier to durable treatment, and ChIP is uniquely suited to uncover how cancer cells rewire their chromatin landscapes to survive. In the case of androgen receptor (AR)-driven prostate cancer, ChIP-seq has shown how alternative AR binding sites emerge in response to therapy, leading to reactivation of growth programs even in the absence of ligand. These adaptations are not random; they often involve pioneer factors like FOXA1 and chromatin remodelers that facilitate new transcriptional circuits. By mapping these events in real time, ChIP enables clinicians to anticipate resistance pathways and design combination therapies to preempt them.

ChIP also intersects with pharmacogenomics. By integrating ChIP-seq with drug response profiles, researchers can predict which transcriptional regulators are most critical to a given cancer subtype and tailor therapies accordingly. For instance, in triple-negative breast cancer (TNBC), the discovery of BRD4-dependent super-enhancers through ChIP-seq has led to clinical trials of BET inhibitors. These compounds, once guided by empirical observation, are now rationally deployed based on ChIP-informed biomarkers. Similarly, the role of transcription factor dependencies in childhood leukemias—such as RUNX1 in t(8;21) AML—has been clarified through ChIP-seq, informing the development of lineage-specific therapies.

As ChIP continues to evolve, so too does its clinical relevance. While traditionally confined to research labs, advances in low-input ChIP, microfluidic platforms, and CUT&Tag are enabling ChIP-like data acquisition from patient biopsies. This has tremendous implications for personalized medicine. Imagine a future where a biopsy not only yields a mutational profile but also a chromatin map—pinpointing which enhancers are active, which silencers are misfiring, and which transcription factors dominate the regulatory landscape. Such data could inform everything from drug choice to prognosis to eligibility for clinical trials.

In viral diseases as well, ChIP has played a crucial role in understanding host-pathogen interactions. In HIV latency research, ChIP-seq has revealed that the integrated viral genome is wrapped in repressive histone marks that prevent transcription. Understanding this silencing mechanism is central to the "shock-and-kill" therapeutic strategy, wherein latent reservoirs are reactivated and cleared. In hepatitis B virus (HBV) research, ChIP-seq has mapped host transcription factor binding to viral cccDNA, shedding light on mechanisms of persistence and reactivation. These examples show that the boundaries between virology, immunology, and epigenomics are increasingly porous—and ChIP sits at the intersection.

The convergence of ChIP with CRISPR technologies further expands its disease relevance. CRISPR interference (CRISPRi) and CRISPR activation (CRISPRa) allow precise modulation of gene expression, and ChIP-seq can confirm whether these manipulations change chromatin states as expected. This creates a virtuous cycle of perturbation and validation, where hypotheses are not only tested but epigenetically verified. Such precision is vital in diseases with complex regulatory architectures, such as those involving noncoding RNAs, enhancers, and long-range chromatin interactions.

Ultimately, the clinical and translational potential of ChIP lies in its unmatched ability to read the genome not just as a static sequence, but as a dynamic, interpretive script. It tells us where the actors—transcription factors, histone modifications, RNA polymerases—enter and exit the stage. It shows which scenes are repressed and which are amplified, which roles are cast and which are silenced. In cancer and other diseases, where the play goes tragically awry, ChIP offers a script doctor's insight—revealing the editorial marks, the miswritten lines, and perhaps even suggesting rewrites that restore order. In doing so, it bridges the bench and the bedside, illuminating not just the biology of disease but the pathways toward its resolution.

26. USING ChIP IN DEVELOPMENTAL AND STEM CELL BIOLOGY

The dance of life begins long before the first heartbeat. It is choreographed in silence, orchestrated by the imperceptible binding of proteins to DNA, and recorded in the chromatin landscapes that change as a single cell transforms into the complexity of a living organism. Chromatin immunoprecipitation (ChIP) has become an indispensable instrument for decoding the molecular symphony that underlies development and stem cell biology. It offers a lens into how gene regulation changes not just between cell types but through time itself—revealing the dynamic interplay between transcription factors, histone modifications, and the regulatory elements that determine cellular identity. In this chapter, we venture into the evolving genome of developing organisms and stem cells, guided by the precision and insight ChIP affords.

The allure of embryonic stem cells lies in their dual capacity: the ability to self-renew indefinitely and the potential to differentiate into virtually any cell type. This remarkable plasticity is maintained by a core transcriptional circuitry involving key regulators such as OCT4, SOX2, and NANOG.

ChIP experiments in embryonic stem cells were among the first to map the genomic targets of these master regulators, revealing extensive networks of autoregulation and feed-forward loops. OCT4 and NANOG do not merely activate genes promoting stemness—they also repress differentiation pathways, safeguarding the ground state of pluripotency. These transcription factors bind to thousands of sites across the genome, often co-occupying enhancers and promoters in tight coordination, hinting at a cooperative model of gene control that resists binary switches in favor of complex modulation.

One of the most striking contributions of ChIP to stem cell biology is the discovery of "bivalent domains"—genomic regions simultaneously marked by histone modifications associated with both activation (H3K4me3) and repression (H3K27me3). These conflicting signals are not signs of confusion but of poised potential. Found primarily in embryonic stem cells, bivalent domains sit atop developmental gene promoters, keeping them transcriptionally silent yet ready to be activated during differentiation. The existence of these dual-marked sites supports a model in which stem cells prime key lineage-specifying genes, keeping them in a state of suspended readiness until the appropriate external or internal cue arrives. As differentiation proceeds, the bivalent nature resolves—H3K27me3 marks are retained or lost depending on the fate of the cell, leading to full repression or activation of these crucial genes.

ChIP has allowed researchers to follow the trajectory of cell fate transitions with unprecedented granularity. By performing ChIP at sequential timepoints during differentiation protocols, scientists have been able to observe the progressive remodeling of chromatin architecture. Enhancers are activated in waves, often marked first by increased H3K4me1 and then by acquisition of H3K27ac, signifying a transition from primed to active states. Transcription factors act in cascades, with early developmental regulators binding to broadly accessible regions and later factors refining the specificity of gene expression. This time-resolved application of ChIP has illuminated key transitions, such as the shift from mesendoderm to endoderm in early embryogenesis or the sequential specification of neuronal subtypes in the developing cortex.

Model organisms have played a pivotal role in validating and expanding our understanding of development through ChIP-based studies. In mice, ChIP has been used to map enhancer usage and transcription factor binding across embryonic days, revealing the genomic underpinnings of morphogenesis and organogenesis. Zebrafish, with their transparent embryos and rapid development, provide another fertile ground for developmental ChIP research. Using ChIP-seq on sorted cell populations from staged embryos, scientists have mapped histone marks and factor binding in specific germ layers and organ progenitors. These efforts have produced chromatin atlases of development, which serve as invaluable references for identifying candidate enhancers, lineage determinants, and the molecular roots of developmental disorders.

Stem cell reprogramming, the process of converting somatic cells back to a pluripotent state, owes much of its mechanistic clarity to ChIP. The forced expression of OCT4,

SOX2, KLF4, and c-MYC (collectively known as OSKM) induces a radical reorganization of the epigenome. ChIP studies have traced the journey of these reprogramming factors as they bind closed chromatin, recruit remodeling complexes, and gradually activate endogenous pluripotency networks. The process is neither immediate nor uniform; cells traverse intermediate states, each with distinct chromatin features. Bivalent marks are re-established on developmental genes, enhancers regain accessibility, and a new landscape of promoter-bound transcription factors emerges. ChIP helps chart these milestones, offering a roadmap of epigenomic conversion that is vital for improving reprogramming efficiency and fidelity.

Yet the application of ChIP in developmental and stem cell biology is not without challenges. The number of cells available from embryonic tissues or rare progenitor populations is often limited, demanding innovations in low-input ChIP methodologies. Techniques such as ChIPmentation and CUT&RUN have addressed this limitation by reducing material requirements and improving signal-to-noise ratios. Sorting specific cell types using fluorescence-activated cell sorting (FACS) adds further specificity but introduces technical hurdles such as fixation optimization and chromatin integrity preservation. Despite these constraints, the field has rapidly expanded, with increasingly refined protocols making it possible to profile histone modifications and factor binding in as few as hundreds or thousands of cells.

The utility of ChIP extends beyond the embryo to adult stem cells and regeneration. In hematopoietic stem cells, ChIP has elucidated how transcription factors like PU.1, GATA2, and RUNX1 establish lineage biases and maintain quiescence. In intestinal crypts, ChIP has shown how Wnt-responsive enhancers are selectively activated in stem and transit-amplifying cells. Even in the brain, where adult neurogenesis persists in restricted niches, ChIP has helped identify enhancers and repressors that govern neuronal fate decisions. Across these tissues, a recurring theme emerges: developmental potential is governed not by static genetic programs but by fluid, context-sensitive epigenetic landscapes that ChIP helps to decode.

Understanding development is not solely a pursuit of curiosity—it bears profound implications for human health. Many congenital disorders stem from misregulation of developmental genes, often through mutations in enhancers or transcription factors. ChIP can help pinpoint the functional impact of such mutations, clarifying whether they alter binding affinity, chromatin state, or transcriptional outcomes. In regenerative medicine, insights from ChIP guide the design of differentiation protocols, ensuring that stem cells adopt desired fates efficiently and safely. The therapeutic promise of organoids, tissue engineering, and cell therapy all rests on our ability to faithfully recapitulate developmental cues—an endeavor that hinges on the chromatin maps ChIP provides.

In its essence, ChIP transforms abstract concepts of gene regulation into tangible landscapes—peaks, valleys, and territories that we can visualize, measure, and manipulate. In the realm of developmental and stem cell biology, these landscapes are not merely maps—they are blueprints. They reveal how a genome unfolds in time and space, how identity is sculpted from potential, and how life orchestrates itself from within. From the pluripotent ground state to terminal differentiation, from the embryo's first axis to the formation of organs, ChIP tracks the molecular choreography with elegance and precision. It allows us not just to witness development, but to understand it—deeply, mechanistically, and with the clarity required to intervene when nature's design falters.

Development is a symphony of genetic instructions, orchestrated with exquisite timing and precision. At its heart lies the regulation of gene expression, where chromatin state and transcription factor binding dictate cellular identity and fate. Chromatin Immunoprecipitation (ChIP) has emerged as a critical technique for capturing the shifting regulatory landscapes that guide cells from pluripotency to terminal differentiation. Nowhere is this more evident than in the field of developmental and stem cell biology, where ChIP offers a window into the dynamic interplay between chromatin modifiers, transcription factors, and the epigenetic marks that shape cell destiny. By mapping where proteins bind across the genome at different stages of development, researchers can chart the regulatory milestones that underlie morphogenesis and organogenesis.

In the early embryo, cells possess remarkable plasticity, with the ability to generate any cell type. This pluripotent state is governed by a network of transcription factors, most notably OCT4, SOX2, and NANOG. These factors form a self-sustaining regulatory circuit that maintains the undifferentiated state, and ChIP studies have been instrumental in defining their binding sites across the genome. OCT4, for instance, binds to thousands of enhancers and promoters, often in conjunction with other pluripotency factors, to activate genes essential for maintaining stemness. By overlaying OCT4 ChIP-seq data with RNA-seq expression profiles, researchers have uncovered direct transcriptional targets and revealed how OCT4 contributes to both gene activation and repression depending on chromatin context.

A hallmark discovery in embryonic stem cells using ChIP was the identification of bivalent chromatin domains—genomic regions simultaneously marked by H3K4me3 (an active mark) and H3K27me3 (a repressive mark). These domains poise developmental genes for rapid activation or stable silencing upon differentiation. ChIP-seq mapping of these histone marks revealed that bivalency is prevalent in genes involved in lineage specification, suggesting a mechanism by which pluripotent cells keep developmental options open. As cells begin to differentiate, bivalent domains resolve into monovalent states, with one mark predominating to either commit to expression or repression. This elegant switch is choreographed in part by polycomb group proteins and Trithorax complexes, whose occupancy can also be tracked by ChIP.

Temporal ChIP-seq has added another dimension to developmental studies, allowing researchers to capture chromatin and transcription factor dynamics over time. In neural differentiation models, for example, the sequential binding of SOX2, followed by NEUROD1 and other neurogenic factors, has been mapped using ChIP. Each transcription factor binds distinct sets of enhancers, initiating cascades of gene expression that push cells along the neural trajectory. Heatmaps of ChIP signal intensities across developmental timepoints reveal waves of enhancer activation and histone mark deposition that accompany lineage commitment. In hematopoiesis, a similarly dynamic choreography of factors like PU.1, GATA1, and RUNX1 has been uncovered through time-resolved ChIP, shedding light on how multipotent progenitors give rise to diverse blood lineages.

One of the most powerful applications of ChIP in development has been the study of induced pluripotent stem cells (iPSCs). When somatic cells are reprogrammed using factors like OCT4, SOX2, KLF4, and c-MYC, ChIP-seq has been used to trace how the chromatin landscape is remodeled to resemble that of embryonic stem cells. Early in reprogramming, the introduced factors bind to a subset of accessible loci, but over time, their occupancy spreads to pluripotency-associated enhancers and promoters, coinciding with activation of endogenous stem cell genes. ChIP has also revealed that many regions remain in a partially reprogrammed state, with incomplete histone mark profiles or aberrant binding patterns. These findings underscore the epigenetic barriers to full reprogramming and have guided strategies to enhance iPSC fidelity.

In model organisms such as zebrafish, mouse, and Drosophila, ChIP has illuminated developmental processes from gastrulation to organogenesis. In zebrafish embryos, ChIP for histone modifications has mapped enhancer usage in real-time during body plan formation, while in mouse embryos, lineage-specific transcription factor binding has been profiled in dissected tissues. These in vivo studies face unique challenges—limited cell numbers, heterogeneity, and the need for precise staging—but advances in low-input and single-cell

ChIP methods have begun to overcome these obstacles. Cell sorting strategies using fluorescent reporters or surface markers enable enrichment of specific populations, while improvements in antibody sensitivity and tagmentation-based protocols enhance signal recovery.

Perhaps one of the most striking insights from developmental ChIP studies is the modularity and reuse of regulatory elements. Enhancers activated in early development may be redeployed in later stages, with different transcription factors occupying the same sequences depending on cellular context. This concept of enhancer repurposing has been demonstrated in heart development, where MEF2 and GATA factors take turns binding to overlapping cis-regulatory elements. ChIP has also revealed long-range chromatin interactions between enhancers and promoters, mediated by looping factors like CTCF and cohesin, which sculpt the 3D genome architecture during tissue specification.

Despite its power, ChIP in developmental contexts is not without limitations. The scarcity of material in early embryos or rare progenitor populations can hinder data quality. Background signal, variability in chromatin accessibility, and antibody specificity all influence interpretation. Nevertheless, newer techniques like CUT&RUN and CUT&Tag, which require fewer cells and yield cleaner profiles, are increasingly being applied to developmental systems. Integration with single-cell transcriptomics and chromatin accessibility assays provides a richer picture of how transcriptional programs unfold over time.

Ultimately, ChIP allows developmental biologists to link the abstract blueprints of fate with the physical signatures left on the genome. It translates transcription factor theory into biochemical maps and transforms static gene lists into dynamic regulatory narratives. By following the footprints of DNA-binding proteins and histone modifiers, scientists uncover how cells navigate the branching roads of differentiation and retain—or relinquish—their identity. As technology advances and the frontier of stem cell biology expands, ChIP will remain a compass, guiding discovery through the epigenetic landscapes that build life from a single cell.

27. COMMON PITFALLS AND TROUBLESHOOTING

The promise of chromatin immunoprecipitation lies in its ability to unlock the hidden architecture of gene regulation, revealing where proteins meet DNA to orchestrate expression. Yet, for many researchers, the path to reliable ChIP data is littered with frustration. Experiments may yield flat signals, noisy backgrounds, or mysteriously irreproducible results, leading even the most seasoned scientists to question their technique, their reagents, or sometimes their scientific intuition. This chapter is a companion for that critical moment—the place between setback and success—offering clarity, practical strategies, and the confidence to interpret and correct what has gone wrong.

The first, and perhaps most frequent, challenge encountered in ChIP is low signal. This often manifests as little or no enrichment at expected binding sites, despite positive controls and careful experimental setup. The culprit may be the antibody. Antibody quality is paramount; even commercially available antibodies may not be validated for ChIP-grade performance. A ChIP-validated antibody must not only bind its target with high specificity but also tolerate the structural disruptions of crosslinking and the harshness of fragmentation. Batch-to-batch variability, poor lot documentation, and inadequate storage can all diminish antibody function. Researchers must ensure rigorous lot validation—ideally testing new batches on known targets or pilot samples—and confirm antibody specificity with western blotting, dot blot, or peptide blocking assays before scaling to ChIP.

Low signal can also stem from suboptimal crosslinking. Crosslinking with formaldehyde is a delicate balance: under-crosslinking leads to loss of protein-DNA complexes during lysis and wash steps, while over-crosslinking hinders epitope recognition and chromatin fragmentation. The standard condition—1% formaldehyde for 10 minutes at room temperature—is often effective but not universal. For proteins that bind transiently or loosely to DNA, shorter fixation times or dual crosslinking with longer-chain reagents like DSG or EGS may be necessary. Conversely, for proteins deeply buried in chromatin or part of large complexes, a more intense fixation may be appropriate, albeit with greater risk to antibody recognition.

Next comes the mechanical heart of ChIP: chromatin fragmentation. Whether using sonication or enzymatic digestion, this step determines the resolution and quality of the resulting data. In sonication, energy input must be tightly controlled—too little, and chromatin remains intact and unusable; too much, and protein-DNA complexes may be destroyed. Variables such as cell type, fixation strength, and volume affect sonication efficiency, and outcomes must be verified with DNA gel electrophoresis or bioanalyzer traces. Optimal fragment sizes are typically between 100 and 500 base pairs, a range that maintains immunoprecipitation efficiency while offering good resolution. For enzymatic shearing (e.g., micrococcal nuclease), digestion time and calcium concentration must be carefully calibrated, as overdigestion can erase the very footprints researchers seek to detect.

High background noise is another common issue. This may appear as significant signal in the input or IgG control samples or an indistinct peak landscape across the genome. Poor washing protocols are often to blame. Washing buffers should balance stringency and preservation of complexes; low-salt washes may fail to remove nonspecifically bound DNA, while high-salt or detergent-rich buffers may destabilize protein-DNA complexes. An empirical approach—adjusting salt concentrations, adding NP-40 or SDS, extending wash times—can help. Including sequential washes with low- and high-stringency buffers is a common solution that preserves signal while reducing background.

Reproducibility, or the lack thereof, looms large in ChIP. Discrepancies between replicates, labs, or time points can cast doubt on entire datasets. Variability in input material—cell confluency, growth conditions, passage number—can subtly shift chromatin landscapes, especially in sensitive cells like stem cells or primary cultures. The amount of starting material, typically 1–10 million cells per IP, must be consistent, as too little DNA can exaggerate noise and bias. Controls are essential: include input DNA, mock IPs (e.g., IgG or beads only), and spike-in standards if possible. The addition of external chromatin standards, such as from Drosophila or yeast, provides a normalization baseline for signal intensity and helps account for technical variability across samples or sequencing batches.

Loss of DNA during ChIP—particularly in low-input scenarios—is a silent but devastating pitfall. It can occur at multiple stages: during lysis, washing, elution, or purification. Low-binding tubes, DNA LoBind plastics, and pre-blocked beads reduce loss from adhesion. Glycogen, linear acrylamide, or carrier RNA can improve recovery during precipitation. DNA purification columns must be chosen with care, as some preferentially bind larger fragments or lose DNA during elution. For low-input ChIP, carrier proteins or DNA-compatible elution buffers can significantly boost recovery without affecting downstream qPCR or sequencing.

Another source of failure lies in protein-specific challenges. Transcription factors that bind DNA weakly, transiently, or in low abundance pose special difficulties. ChIP of such factors may require optimization of crosslinking, increased cell input, or sequential ChIP (re-ChIP) to enrich target interactions. Histone modifications, though abundant, present their own issues. Many histone marks occur in broad domains rather than sharp peaks, requiring specialized peak callers and interpretation. Moreover, the sheer number of available histone antibodies—many of dubious ChIP specificity—demands careful validation. Dot blots against modified peptides, peptide competition assays, or orthogonal validation via CUT&RUN or CUT&Tag offer reassurance.

Even with careful preparation, pitfalls arise in data analysis. Poor peak calling can stem from inadequate control datasets or unfiltered low-quality reads. Tools like MACS2 perform best with matched inputs and carefully trimmed, aligned reads. Adapter contamination, low-complexity reads, and PCR duplicates must be filtered out. False positives can arise from mapping artifacts or repetitive elements; blacklist regions from ENCODE help flag such genomic trouble zones. For comparative studies, normalization methods like RPKM, spike-in scaling, or DESeq2 should be used to control for sequencing depth and library complexity.

To help navigate these pitfalls, visual tools can be a lifesaver. Genome browsers like IGV or UCSC allow immediate inspection of peaks, coverage, and artifacts. If an expected target shows no signal or if background is oddly elevated, visualization often reveals the pattern. Peaks that appear in all samples, including negative controls, are likely technical artifacts. Conversely, peaks found only in the ChIP sample at biologically plausible sites reinforce confidence in success.

For researchers who find themselves stuck, structured troubleshooting guides offer a diagnostic lifeline. These are often presented as flowcharts or tables: "No signal? Check antibody > Check crosslinking > Check chromatin fragmentation." Decision trees that walk the user through symptom, possible cause, and corrective action reduce confusion and streamline recovery. Examples drawn from published studies—where researchers overcame technical setbacks to publish high-impact data—can be powerful affirmations that success is still within reach.

Common errors also yield teachable moments. A lab might discover, after months of failed ChIP, that their antibody was shipped on dry ice but stored at room temperature for a week. Another may realize their sonicator, despite humming reassuringly, had a broken probe. Yet another team might trace background noise to contaminated beads or faulty blocking agents. These stories, shared across conferences, forums, and publications, reinforce the value of meticulous documentation, routine validation, and the humility to recheck every assumption.

As ChIP evolves—with newer techniques like ChIP-exo, CUT&RUN, and CUT&Tag offering simpler workflows—many pitfalls are being circumvented. But even these streamlined methods are not immune to error. Enzyme activity, tagmentation efficiency, and adapter ligation must be verified. And for those committed to classical ChIP, mastering the troubleshooting process is a rite of passage. It transforms frustration into fluency, confusion into control.

Ultimately, the best troubleshooting strategy is prevention. This includes rigorous experimental planning, complete protocols, standard operating procedures, and adequate training for all lab members. Pilot studies, positive and negative controls, and redundant validations all reduce the risk of misinterpretation. A single well-designed ChIP experiment, executed carefully and analyzed critically, provides more value than a dozen poorly controlled ones. ChIP is not a black box—it's a window into the living genome. And with the right mindset and tools, every pitfall becomes a stepping stone toward mastery.

28. EXPERIMENTAL DESIGN AND REPRODUCIBILITY BEST PRACTICES

Sound experimental design is the backbone of every meaningful ChIP study, serving as the silent architect of all downstream success. While the power of ChIP lies in its ability to illuminate the physical association between proteins and DNA, this power can easily be lost—or worse, misrepresented—if the initial planning is flawed. This chapter, therefore, is not about flashy results or novel discoveries; it is about the discipline of design, the rigor of reproducibility, and the unglamorous but vital scaffolding that holds scientific truth aloft. A poorly controlled ChIP experiment may produce what looks like a compelling peak, but in reality, it could be a mirage conjured by noise, contamination, or careless planning. The difference between signal and artifact often lies not in the antibodies or sequencers used, but in the wisdom of the design.

Controls are the first bulwark against false discovery. The inclusion of input controls, IgG controls, and mock immunoprecipitations is non-negotiable. The input serves as the background—genomic DNA that has undergone all steps except for the immunoprecipitation—and without it, it's impossible to identify real enrichment from the surrounding genomic landscape. The IgG control offers an additional layer, reflecting the level of non-specific binding that might occur in the absence of a specific antibody. Each control sharpens the silhouette of the real signal, much like negative space gives shape to a painting. Some may be tempted to skip these steps in the name of efficiency or cost savings, but the expense of publishing—or worse, acting on—erroneous data is far greater.

Replicates, too, are not just academic niceties. Biological replicates demonstrate that an observed effect is not a one-off event or an accident of cell culture or sample preparation. They ensure that what's being seen is truly characteristic of the system under study. Technical replicates, meanwhile, can reveal the subtle quirks and noise introduced by the process itself—variability in sonication, antibody performance, or library preparation. Both types of replications must be judiciously planned. A single replicate is a hint; two are suggestive; three begin to approach confidence. Yet not all replicates are created equal. Performing three experiments from the same culture split into tubes is not equivalent to harvesting three truly independent biological samples. The depth and diversity of replication determine the reliability of interpretation.

Statistical power in ChIP experiments is often overlooked, yet it is as important here as in clinical trials or gene expression studies. Power calculations can guide the number of replicates needed, the sequencing depth required, and the sensitivity of detection. Sparse data is easy to over-interpret; noise has a way of masquerading as significance in the absence of context. Batch effects—those insidious variables introduced by day, operator, reagent lot, or equipment—can quietly confound results. Designing experiments to account for and mitigate these effects, by randomizing samples across batches and including batch-specific controls, is essential. Blinded analysis, though more common in clinical disciplines, has merit in ChIP studies as well. Analysts unaware of sample identity are less likely to subconsciously cherry-pick or rationalize anomalies.

Once data has been generated, reproducibility becomes a multi-headed challenge. Computational reproducibility begins with metadata: the complete and precise record of every step, every version, every parameter. Which antibody lot was used? What sonicator settings? What genome build? The answers to these questions may seem mundane in the moment but become pivotal months or years later, especially when inconsistencies arise. Version tracking software, such as Git or commercial lab notebooks with audit trails, ensures that analytical pipelines are not black boxes. Every alignment parameter, every peak caller threshold, every filtering step must be recorded and, ideally, shared. Transparency is not optional; it is the price of credibility.

The publication of ChIP data now comes with certain expectations. Journals increasingly require that raw and processed data be deposited in public repositories such as GEO or SRA. But beyond availability, the structure and quality of the data matter. Standards developed by consortia like ENCODE and modENCODE provide benchmarks. These include criteria for signal-to-noise ratio, cross-correlation metrics, and peak reproducibility. Meeting or exceeding these thresholds is no longer just best practice—it is often a prerequisite for peer-reviewed publication. These guidelines also encourage rigorous documentation, including detailed protocol metadata, sample annotations, and descriptions of controls and replicates.

Even before publication, the discipline of annotation improves the clarity and utility of a study. Using standardized vocabularies for tissue type, cell line, developmental stage, or treatment condition makes datasets more discoverable and interoperable. Employing universally accepted formats such as BED, BigWig, and BAM ensures that results can be viewed and validated in genome browsers without friction. Labeling is not an afterthought—it is the architecture that allows data to be integrated and built upon. The proliferation of online repositories and data portals has made it easier than ever to share ChIP-seq data, but also increased the burden to make sure it is presented in a coherent and interpretable form.

Transparent documentation and reproducibility also serve a broader mission: they anchor trust in science itself. In an era where scientific credibility is often under public scrutiny, the ability to replicate findings is not just a methodological concern—it is an ethical one. Transparency protects against both honest error and dishonest manipulation. It provides a framework in which findings can be interrogated, validated, or refuted. For young researchers, this ethos should be ingrained from the start: it is not enough to get interesting results; one must be able to defend them, explain them, and allow others to test them. The beauty of science lies not only in discovery, but in the durability of that discovery under scrutiny.

The ENCODE consortium provides a model of what high-quality ChIP-seq data generation should look like. Their meticulous standards, extensive metadata, and open data policy have turned their dataset into a global resource.

Following their lead, newer projects now publish not just data, but the entire provenance of that data—tools used, parameters set, and statistical measures applied. As datasets grow more complex, including multi-omic integration with RNA-seq, ATAC-seq, or Hi-C, the demand for coherent, reproducible pipelines grows as well. Many labs now utilize workflow management systems like Snakemake, Nextflow, or CWL to codify their analyses, making them both portable and reproducible. These tools, combined with containerized environments like Docker, can freeze the computational context, ensuring that years from now the same code yields the same results.

Randomization in sample processing helps to guard against bias introduced by unconscious expectations or procedural artifacts. For example, if all treated samples are prepared on a Monday and controls on a Friday, differences observed may reflect handling discrepancies rather than biology. True randomization distributes these confounders evenly, making real differences stand out more clearly. Similarly, when data interpretation is performed blind to sample identity, the temptation to favor expected outcomes is reduced. In ChIP studies involving differential binding between conditions—say, treated versus untreated cells—this approach becomes particularly important. Decisions about thresholding, noise filtering, or peak significance can shift subtly depending on what the analyst hopes to see.

Yet, even with the best planning and tools, reproducibility is not a checkbox to be ticked—it is a culture. It must be embedded in the way teams collaborate, in the way data is discussed, and in the way results are communicated.

Reproducibility should be built into the lab's DNA, reinforced at lab meetings, encoded into standard operating procedures, and demanded in manuscript preparation. The benefit of this culture becomes most evident in times of crisis—when a new antibody lot fails, when sequencer output plummets, or when a critical result is challenged. Labs steeped in reproducibility don't scramble to reconstruct what they did; they simply consult the record.

In sum, experimental design in ChIP is not merely a matter of protocol—it is a matter of philosophy. To conduct ChIP without rigor is to risk drawing maps to phantom cities. But with disciplined design, thorough documentation, thoughtful control selection, and an unwavering commitment to reproducibility, ChIP becomes a compass that points not just to signal, but to truth. The responsibility lies with every scientist not just to uncover patterns in the genome, but to do so in a way that others can follow, verify, and build upon. This chapter, therefore, is a manifesto as much as a manual—a reminder that great science is as much about how we ask as what we find.

Yet at the heart of all sound experimental planning lies a mindset—not just of meticulousness, but of humility. It is the understanding that the biological world is far more complex and entangled than any single protocol can capture. ChIP may be a powerful technique, but it becomes transformative only when used with scientific restraint and intellectual rigor. The overconfidence of early researchers who assumed one replicate would suffice has been replaced with the awareness that biology thrives on variability, and true patterns emerge only when noise is acknowledged and accounted for.

This is why robust experimental design begins not at the bench but at the whiteboard. A good ChIP study is visualized before it's realized. Inputs, outputs, and checkpoints are drawn like battle plans—each replicate accounted for, each possible source of error anticipated. Biological replicates should be as independent as possible, originating from different cultures or animals, reflecting true biological diversity rather than technical repetition. Technical replicates, meanwhile, help diagnose protocol consistency. Combining both in thoughtful balance is no small feat; it's a hallmark of a researcher who doesn't merely seek to confirm hypotheses but to confront them with data that could dismantle or elevate them.

Controls are not a formality—they are the bedrock. A ChIP without a proper input control is like a photograph without contrast. Input DNA accounts for background shearing patterns, copy number variations, and technical artifacts. Mock IPs, using IgG or beads alone, help detect nonspecific interactions, while spike-ins such as Drosophila chromatin allow for normalization across samples, enabling true comparative analysis. These tools aren't just technical—they're philosophical. They remind us that signal without context is meaningless.

Batch effects can unravel even the most carefully crafted experiments. A sample processed on Monday may differ from one processed Friday, not because of biology, but because of the weather, the technician's mood, or the bottle of buffer that was almost empty. Randomization helps mitigate this—by processing replicates across different days, instruments, and personnel, we embed variation into the experiment intentionally, not accidentally. Paired with blind analysis, where the analyst doesn't know which group a sample belongs to, we build protection against our own biases—those invisible hands that can tip a histogram or nudge a peak just far enough to confirm our theory.

Statistical power must also be considered early, not as a post-hoc rationalization. Underpowered studies may show tantalizing trends that crumble under scrutiny. Using power analysis to estimate the number of replicates needed to detect a given effect size ensures we're not wasting time chasing shadows. This is especially vital when working with expensive or limited samples—developmental tissue, primary cells, or rare patient biopsies. The cost of one additional replicate pales in comparison to the cost of drawing conclusions from a false positive.

Once the experiment is underway, documentation becomes the silent partner of reproducibility. Metadata—the who, what, when, and how of each sample—should be captured in structured formats, ideally using lab information management systems (LIMS) or standardized spreadsheets. Sample barcoding, reagent lot tracking, and software version logging are not obsessive compulsions but essential practices. As datasets grow larger and analyses more complex, being able to trace a result back to its experimental root is the difference between validation and vapor.

Modern standards from consortia like ENCODE and modENCODE have crystallized these principles into concrete guidelines. These bodies have helped define minimal criteria for replicate number, sequencing depth, signal-to-noise ratios, and peak reproducibility. Their white papers, though often dense, are goldmines of wisdom distilled from thousands of failed and successful ChIP experiments. Adhering to these guidelines not only prepares researchers for publication but ensures that their data can stand the test of time, scrutiny, and reanalysis.

Tools have emerged to assist in this quest for transparency and fidelity. Platforms like GEO and SRA enable public deposition of raw and processed data, while TrackHub and UCSC Genome Browser offer visual portals for reviewers and peers to inspect results firsthand. Version control systems like Git can track code and analysis evolution, creating a living record of every decision made from FASTQ to figure. These practices don't just support reproducibility—they transform scientific publications into reusable, dynamic resources.

At its core, reproducibility is not a burden but a gift to the future. A well-documented, well-powered, well-randomized ChIP experiment becomes a legacy—a dataset that other scientists can use, reanalyze, reinterpret, and extend. It becomes part of the invisible scaffolding of knowledge that holds up new discoveries, that saves others from repeating mistakes, that accelerates progress.

And perhaps most importantly, reproducibility is trust made tangible. In a world where scientific skepticism sometimes overshadows curiosity, and where data can be politicized or misused, the reproducible experiment is a bulwark. It says: here is the method, here is the evidence, and here is the path we took. You can walk it too.

By embracing these best practices—not as checkboxes, but as a culture—we transform ChIP from a technique into a testimony. A testimony to rigor. A testimony to truth. And a testimony to a scientific community that values what is real over what is easy.

29. ETHICS, DATA SHARING, AND REPRODUCIBILITY IN EPIGENOMIC RESEARCH

In the unfolding era of epigenomics, where every peak in a ChIP-seq dataset might hold clues to disease, identity, or inheritance, the sheer power of the data demands a commensurate responsibility. Chromatin immunoprecipitation, once a niche technique, now generates profiles that map the intimate dance between transcription factors and our genome, revealing when and where genes awaken, which regions remain silent, and how that orchestration varies between health and disease. But with this power comes the growing imperative to examine not just how we generate knowledge, but how we treat the people, systems, and principles entwined with it. In the search for understanding, science can no longer afford to treat ethics, transparency, or reproducibility as afterthoughts. Instead, they must be the scaffolding on which all epigenomic inquiry is built.

Every human ChIP experiment begins not at the bench, but with a promise—often silent, but binding. Whether collecting blood from cancer patients or deriving induced pluripotent stem cells from donor fibroblasts, the foundation lies in informed consent. It's not enough to merely explain the basics of sample use; researchers must clearly articulate how ChIP data, though non-coding, can carry implications far beyond what was once imagined. Enhancer maps can indicate cell types and disease risks; binding patterns may reveal the presence or absence of specific genetic variants. And while raw ChIP-seq reads might not scream "identity" the way exome sequences do, their interpretation in aggregate could, under the lens of modern algorithms, signal far more. The evolving landscape of data science, especially when merged with machine learning, raises the stakes on what constitutes personally identifiable information.

This complexity demands precision and honesty in the consent process. Participants deserve to know not only how their samples will be used today, but how data derived from them may be shared globally, integrated with other modalities, or even reinterpreted in ways that transcend the original study aims. Broad consent models, often used in large-scale initiatives like ENCODE or GTEx, must walk the tightrope between scientific freedom and individual autonomy. The era of "trust us, we're scientists" is over. Participants must become collaborators, not just contributors.

And then comes the matter of data sharing. As the ethos of open science sweeps across disciplines, the need for publicly accessible epigenomic datasets grows. But here too, nuance reigns. Platforms like the Gene Expression Omnibus (GEO) and the European Nucleotide Archive (ENA) are invaluable repositories, enabling others to reanalyze, reinterpret, and build upon existing work. Yet sharing isn't simply a matter of uploading FASTQ files. Each dataset must be accompanied by rigorously curated metadata: sample origin, treatment conditions, antibody information, sequencing depth, library preparation methods, and quality control metrics. Without such context, the raw data become meaningless—peaks without anchors, signals without stories.

The FAIR principles—Findable, Accessible, Interoperable, and Reusable—offer a blueprint for responsible sharing. Making a ChIP-seq dataset "findable" means using standardized vocabularies and persistent identifiers; making it "accessible" implies both technical availability and clarity of licensing. Interoperability ensures datasets can be integrated with other omics modalities, and reusability requires that data be so well-documented and quality-checked that future researchers, even decades from now, can trust its origins and utility. In the context of epigenomics, where enhancer states shift subtly and data dimensionality is immense, FAIRness is not just a virtue—it's a necessity.

Yet even the most ethically sourced and transparently shared dataset can fall short if it's not reproducible. In a world where computational pipelines stretch into dozens of steps—from adapter trimming and alignment to peak calling, differential analysis, and visualization—the margin for error is vast. Batch effects, software version discrepancies, and subtle parameter tweaks can dramatically shift results. A ChIP-seq analysis run today may not yield the same peaks when repeated a month later on a different system. This is where reproducibility becomes not a buzzword, but a lifeline.

Initiatives such as the ENCODE Consortium, modENCODE, and the NIH Roadmap Epigenomics Project have set gold standards, not only for experimental protocols but also for bioinformatic pipelines. Their recommendations go beyond mere best practices; they define what it means to produce trustworthy epigenomic science. Datasets must include replicate concordance metrics, input controls, and clear thresholds for peak calling. Ideally, each dataset's processing pipeline should be containerized or scripted, allowing others to reproduce the exact analysis environment using tools like Docker or Nextflow. Version control systems such as Git, when used to track changes in analysis code, ensure that revisions are transparent, not lost in the fog of informal tweaking.

This focus on reproducibility dovetails with deeper questions about scientific culture. As pressure mounts to publish rapidly, researchers may be tempted to skip replicates, gloss over failed experiments, or cherry-pick the most compelling peaks. But ChIP is unforgiving. Antibody variability, chromatin accessibility differences, and sample degradation can skew results in subtle ways. The path forward requires not just technical vigilance, but a cultural shift—one in which publishing a failed ChIP is as valued as a successful one, provided the failure teaches something real. Journals and funding bodies must reward transparency over polish.

Ethics in epigenomic research also extends into novel territories—especially where artificial intelligence enters the scene. Deep learning algorithms now predict transcription factor binding from raw DNA sequence, integrating ChIP-seq data from hundreds of experiments to train their models. These tools can generate synthetic enhancer maps, impute missing datasets, or even hypothesize how mutations reshape regulatory landscapes. But what happens when such predictions are wrong? When a model trained on one ancestry fails to generalize to another? When imputation obscures genuine biological diversity under the illusion of algorithmic neatness?

The developers of such tools carry ethical obligations, just like wet-lab scientists. Training datasets must be diverse and well-documented. Validation procedures must be transparent. The models should not be treated as oracles, but as starting points for further biological verification. Interpretability—knowing why an algorithm made a given prediction—is as crucial as accuracy. And where such models are deployed in medical contexts, for example to identify putative regulatory mutations in cancer genomes, the stakes rise higher still.

Finally, there is the human element. Behind every dataset lies a team—researchers who pipetted samples, ran sequencers, coded pipelines, and annotated peaks. Their contributions must be fairly acknowledged, their efforts not erased by the simplicity of a downloadable file. And behind them stand the participants—those who donated tissue, consented to use, and trusted that their biological essence would be handled with care. In honoring them, we honor the integrity of science itself.

As the field of ChIP and epigenomic research marches forward—scaling to single-cell levels, integrating with spatial transcriptomics, coupling with proteomics and CRISPR screens—it must remain rooted in responsibility. Ethics is not a barrier to progress but its guarantor. Data sharing is not a threat to credit, but an amplifier of impact. Reproducibility is not a bureaucratic checkbox, but the lifeblood of truth. This chapter serves not only as a guide to best practices, but as a call to action. To be a ChIP researcher today is to be a steward of the genome, a narrator of chromatin's silent stories, and a custodian of trust. Science, to endure, must be not only brilliant—but just.

30. INNOVATION, AUTOMATION, AND AI INTEGRATION

Chromatin Immunoprecipitation (ChIP) has long stood as a cornerstone in the exploration of protein-DNA interactions, helping scientists map the complex regulatory architecture of the genome. As this technique has matured, so too has its potential, now poised on the edge of a dramatic transformation powered by emerging technologies, automation, and artificial intelligence. This final chapter is not a conclusion but an invitation—an open door to what lies ahead. As the scientific community steps forward into the era of high-throughput epigenomics, the future of ChIP beckons with promise, precision, and profound possibilities.

The foundation of tomorrow's ChIP rests in the shrinking of technology and the amplification of power. Microfluidic ChIP-on-a-chip platforms are reimagining the very infrastructure of the experiment. Rather than relying on bulky bench-top equipment and large volumes, researchers are moving toward miniaturized systems where microdroplets flow through engineered channels, enabling reactions to take place in tightly controlled environments. These microfluidic platforms drastically reduce reagent consumption and sample input, while accelerating reaction times and improving consistency. A single device can now perform hundreds of parallel ChIP reactions with minimal human intervention, thereby paving the way for personalized diagnostics and rapid response research.

Alongside this miniaturization comes a revolution in throughput. Automated liquid handling robots now populate the modern epigenomics laboratory. These robotic arms no longer serve merely to pipette; they orchestrate entire experimental workflows. From chromatin shearing to immunoprecipitation and library preparation, automation ensures reproducibility at a scale that manual processing could never hope to achieve. Variability between operators—a long-standing concern in ChIP—is increasingly being eliminated as labs embrace standard operating procedures encoded in robotic logic. Furthermore, integration of barcoding and tracking software enables real-time monitoring of experimental status, reagent use, and data capture, enhancing laboratory efficiency and compliance.

However, physical tools alone are not sufficient to harness the vast data streams emerging from advanced ChIP techniques. Enter artificial intelligence. AI algorithms are rapidly reshaping the interpretation of ChIP data, offering insights that once took months to derive—now available within hours. Deep learning models trained on vast databases of histone marks, transcription factor binding profiles, and chromatin accessibility patterns are now capable of predicting potential binding sites in silico, guiding researchers toward experiments that are more targeted and hypothesis-driven. Tools such as convolutional neural networks (CNNs) and recurrent neural networks (RNNs) can uncover sequence motifs, enhancer-promoter interactions, and context-specific epigenetic signatures with remarkable accuracy.

One of the most exciting developments is the convergence of single-cell technologies. Whereas traditional ChIP experiments average signals across millions of cells, masking cellular heterogeneity, the future lies in the ability to resolve epigenetic landscapes at the level of individual cells. Single-cell ChIP, while currently challenging due to limitations in DNA recovery and amplification, is being rapidly propelled forward by nanodroplet-based systems and highly sensitive barcoded sequencing strategies. But more than just single-cell ChIP, the real paradigm shift is in multi-omics—integrating single-cell ChIP with RNA-seq, ATAC-seq, and even spatial transcriptomics. This synthesis will allow researchers to simultaneously capture transcriptional activity, chromatin accessibility, and histone modifications within the same cell, constructing a dynamic, multidimensional portrait of gene regulation in real time.

As if that weren't enough, machine learning is beginning to penetrate not just the interpretation, but the experimental design itself. Generative adversarial networks (GANs) and transformer-based models such as those inspired by GPT architectures are now being employed to simulate enhancer activity across genomes, even those poorly annotated or from non-model organisms. These generative models can suggest novel binding motifs, predict cooperative transcription factor interactions, and infer regulatory networks from sparse or noisy data. They are particularly valuable in developmental and cancer biology, where epigenetic shifts occur rapidly and in unpredictable ways. By simulating "what-if" scenarios—such as mutations, knockdowns, or drug treatments—these AI engines allow researchers to run virtual experiments before committing precious resources at the bench.

ChIP is also moving closer to the clinic. As protocols become more sensitive and streamlined, the prospect of performing ChIP on biopsy samples, circulating tumor cells, or even liquid biopsies such as blood plasma is gaining momentum. Miniaturized ChIP combined with real-time data interpretation could provide physicians with epigenomic snapshots of a patient's tumor, immune status, or response to therapy. This opens the door to personalized epigenomics—an era where patients receive treatments based on their unique chromatin landscapes. Moreover, AI-driven predictive models can help forecast disease progression, identify early epigenetic markers of resistance, or flag patients at risk based on subtle regulatory shifts long before clinical symptoms appear.

None of this would be possible without open science. The future of ChIP depends not only on technology but on the community's willingness to share data, protocols, and discoveries. Cloud computing platforms now allow researchers from around the world to access, process, and analyze terabytes of epigenomic data without needing local high-performance clusters. Platforms such as the Gene Expression Omnibus (GEO), ENCODE, and the European Nucleotide Archive (ENA) are more than databases—they are living ecosystems of discovery. AI tools are being embedded directly into these repositories, allowing users to generate hypotheses, design experiments, and test models without leaving the digital environment. Metadata standards, version tracking, and containerized pipelines ensure that results are reproducible, scalable, and transparent.

Still, as with all powerful technologies, the evolution of ChIP comes with ethical responsibility. AI-generated interpretations must be transparent and explainable. Clinical ChIP applications must safeguard patient privacy and data integrity. As researchers, it is not enough to pursue knowledge—we must steward it with integrity. FAIR principles (Findable, Accessible, Interoperable, Reusable) must be the norm, not the exception. Efforts to democratize ChIP, through low-cost protocols, community standards, and accessible training resources, will help bridge the gap between elite research centers and emerging global labs.

So, where does this leave us? At the dawn of a new age for ChIP. No longer confined to a niche within molecular biology, ChIP is becoming a central pillar of systems biology, precision medicine, and computational epigenetics. What was once an arduous, time-consuming technique is now a high-throughput, AI-augmented window into the regulation of life itself. As we close this book, we open a broader vista—one in which you, the reader, are no longer just a student of ChIP, but a potential innovator, ready to contribute to its ongoing transformation. Whether your path lies in the wet lab, in the clinic, or in the code of deep learning architectures, the future of ChIP awaits your imagination, your questions, and your courage.

This is not the end. It is the beginning of something extraordinary.

www.ingramcontent.com/pod-product-compliance
Lightning Source LLC
LaVergne TN
LVHW052322200926
848481LV00002B/641
* 9 7 9 8 2 3 1 2 5 9 3 7 3 *